NANOMEDICINE

NANOMEDICINE

Dr. Aparna Bhattacharya

RAJAT PUBLICATIONS
New Delhi

Nanomedicine
by Dr. Aparna Bhattacharya

This edition is published by Rajat Publications
4740/23, Ansari Road Daryaganj,
New Delhi - 110002

ISBN: 978-81-7880-371-5

Edition 2025

Printed and bound in India

RAJAT
PUBLICATIONS

4740/23, Ansari Road, Daryaganj,
New Delhi- 110002 (India)
Tel:- +91-11-23267924
Mob:- +91-9811484973, +91-9811105389
E-mail: rajat_publication@yahoo.com

Preface

Nanomedicine, (the engineering of tiny machines) an offshoot of nanotechnology, refers to the prevention and treatment of disease in the human body. This discipline is in its infancy. It has the potential to change medical science dramatically in the 21 century.

The most elementary nanomedical devices will be used to diagnose illness. Chemical tests exist for this purpose; nanomachines could be employed to monitor the internal chemistry of the body. Mobile nanorobots, equipped with wireless transmitters, might circulate in the blood and lymph systems, and send out warnings when chemical imbalances occur or worsen. Similar fixed nanomachines could be planted in the nervous system to monitor pulse, brain wave activity, and other functions.

Aparna Bhattacharya

Contents

1

Introduction

WHAT IS NANOMEDICINE?

Nanomedicine is the application of nanotechnology to health. It exploits the novel physical, chemical and biological properties of materials at the nanometer range. Nanomedicine has potential impact on the prevention, early and reliable diagnosis and treatment of diseases. Techniques towards ultra-high spatial resolution, molecular resolution and ultra-high sensitivity will provide a better understanding of the cell's complex "machinery" in basic research. The resulting progress should pave the way to more innovative and powerful in-vivo diagnostics tools.

The aim of nanomedicine may be broadly defined as the comprehensive monitoring, repair and improvement of all human biological systems, working from the molecular level using engineered devices and nanostructures to achieve medical benefit. In general terms nanotechnology will have great impact on the methodologies.

The key elements of nanotechnology applied to nanomedicine are:

- The use of analytical tools and devices to bring a better understanding of the molecular basis of disease, patient predisposition and response to therapy, and to allow imaging at the molecular, cellular and patient levels.
- The design of nano-sized multifunctional therapeutics and drug delivery systems to yield more effective therapies.

Nanomedical developments range from nanoparticles for molecular diagnostics, imaging and therapy to integrated medical nanosystems, which may perform complex repair actions at the cellular level inside the body in the future. The most important medical areas where nanotechnology will have a great impact as identified by the European Science Foundation and the European Union are:

NANODIAGNOSTICS

The main goal of nanodiagnostics is to identify diseases at a very early stage at the level of a single cell. Nanotechnology can provide tools for better sensitivity, specificity and reliability. It also offers the possibility to take different measurements in parallel or to integrate several analytical steps from sample preparation to detection into a single miniaturized device. The use of nanoelectronics will improve the sensitivity of sensors based on already established methods. Nanotechnology will improve the microscopic and spectroscopic techniques to achieve ultra-high spatial resolution, molecular resolution and ultra-high sensitivity which will provide a better understanding of the cell's complex mechanisms in basic research. Molecular imaging, aims to create highly sensitive, highly reliable detection agents that can also deliver and monitor therapy. The find, fight and follow concept of early diagnosis, therapy and therapy control will take a new turn with developments in nanotechnology.

Targeted Drug Delivery and Controlled Release

The drug delivery systems enabled by nanotechnology aims to target selected cells or receptors in the body. Nanoformulations which make use of enlarged surface/volume ratio for enhanced reactivity and nanoparticles that can be used as drug carriers will improve the present targeted delivery systems reducing the costs and increasing the patient acceptance. When a drug is suitably encapsulated, in nanoparticulate form, it can be delivered to the appropriate site, released in a controlled way and protected from undergoing premature degradation. These kinds of controlled release techniques enabled by nanotechnology will have less side effects and high efficiency which can be successfully used for the treatment of cancer and wide range of other diseases. Convergence of electronics and controlled releasing techniques are paving way to a new kind of drug delivering technique which has the ability to release drugs on demand. A future vision is that nanoparticles will carry therapeutic payloads or genetic content into diseased cells, minimizing side effects as the nanoparticles will only become active upon reaching their ultimate destination. They may even check for over dosage before becoming active, thus preventing drug released poisoning.

Regenerative Medicine

Regenerative medicine aims to work with the body's own repair mechanisms to prevent and treat disabling chronic diseases such as diabetes, osteoarthritis, and degenerative disorders of the cardiovascular and central nervous system and to help victims of disabling injuries. Nanotechnology has established a cellular and molecular basis for the development of innovative disease-modifying therapies for in-situ tissue regeneration and repair, requiring only minimally invasive surgery. Nanotechnology will help the future therapeutic methods which will be designed to rectify chronic conditions using the body's own healing mechanisms. Nanotechnology can play a pivotal role in the development of cost-effective therapies for in-situ tissue regeneration. The basic elements of importance in this new 'nanobiomimetic' strategy are intelligent biomaterials, bioactive signalling molecules, and cells. The biomaterials are designed to respond to the changes in the immediate environment stimulating regenerative events at molecular level. The sequential signalling triggers the regenerative events at the cellular level which is necessary for the fabrication and repair of cells. Regenerative medicine also aims to effectively exploit the enormous self-repair potential that has been observed in adult stem cells. Nanotechnology will help to identify signalling systems, in order to leverage the self-healing potential of endogenous adult stem cells and to develop efficient targeting systems for stem cell therapies.

2

The Prospect of Nanomedicine

Nanotechnology is a fascinating science for many scientists as it offers them many challenges. One such challenge is Nanorobots, which once thought to be a fantasy has come into reality now. The proposed application of nanorobots can range from common cold to dreadful disease like cancer. Some such examples can be Pharmacyte, Respirocyte, Microbivores, Chromallocyte and many more. The study of nanorobots has lead to the field of Nanomedicine. Nanomedicine offers the prospect of powerful new tools for the treatment of human diseases and the improvement of human biological systems.

The present era of Nanotechnology has reached to a stage where scientists are able to develop programmable and externally controllable complex machines that are built at molecular level which can work inside the patient's body. Nanotechnology will enable engineers to construct sophisticated nanorobots that can navigate the human body, transport important molecules, manipulate microscopic objects and communicate with physicians by way of miniature sensors, motors, manipulators, power generators and molecular-scale computers. The idea to build a nanorobot comes from the fact that the body's natural nanodevices; the neutrophiles, lymphocytes and white blood cells constantly rove about the body, repairing damaged tissues, attacking and eating invading microorganisms, and sweeping up foreign particles for various organs to break down or excrete.

NANOMEDICINE: A TINY DOSE FOR HEALTH

One of the most promising applications of nanotechnology, known as nanomedicine, involves the development of nanoscale tools and machines designed to monitor health, deliver drugs, cure diseases, and repair damaged tissues, all within the molecular factories of living cells and organelles. The NIH Roadmap for Medical Research—the agency's master plan to accelerate the pace of discovery and speed the application of new knowledge to biomedical prevention strategies, diagnostics, and treatments—contains a significant nanomedicine initiative that will begin with the establishment of 3-4 Nanomedicine Development Centres. These multidisciplinary facilities will serve as the intellectual and technological centerpiece of the endeavor. Funding for the centres of $6 million per year.

Today, the initiative's long-term goals sound like scenarios straight out of Isaac Asimov's Fantastic Voyage. nanobots that can search out and destroy cancer cells before they can form tumors. Nanomachines that can remove and replace broken parts of cells, molecule-sized implanted pumps that can deliver precisely targeted doses of drugs when and where they're needed. Even "smart" nanosensors that can detect pathology or perturbation in any or every cell in the body, and instantly communicate that information to doctors. Science fiction may soon become science fact—these and many other nanomedicine innovations are currently in development, and the NIH predicts that its nanomedicine initiative will start yielding medical benefits in as soon as 10 years. Roco also foresees that fully half of all drug discovery and delivery technology will be based on nanotechnology by 2015.

Experts predict that nanosensors will also provide significantly improved tools to determine both internal and external exposures in real time, assess risk, link exposure to disease etiology, characterize gene-environment interactions, and ultimately improve public health. The NIEHS, through its extramural grants and Superfund Basic Research Programme, is funding the research and development behind many of these expected innovations.

This is quite a surprising conception, since few individuals are aware of this new science, as the entire field has received little media or press coverage and is a well kept secret according to Regis. For some individuals the unlimited potential of nanotechnology is to good to be true, or at best, science fiction with cynicism and skepticism included. The technology is so radical, so powerful, that it will be hard to accept. The true believer of nanotechnology is highly ethnocentric in his or her vision.

At the forefront of nanotechnology the potentials are rapidly being realised. The three areas of Biotechnology, Biometric Chemistry, and Atomic Positioning will be the first directions of growth in the field of nanotechnology according to many scientists. Even nanomedicine is currently available in the modern medical clinic. According to David Voss gene therapy is just the start of nanotechnology in the medical field. He along with Mauro Ferrari at Ohio State University are utilizing nanomedicine with state-of-the-art technology to combat genetic diseases. This small vanguard of medical researchers are discovering nanotechniques never used before. For instance, Ferrari is experimenting with nanodelivery of new cells into the human body to replace existing diseased or damaged cells. Other researchers are attempting to construct nanobombs to fight cancer. Still other medical researchers argue that the upcoming change in medicine will offer an unlimited number of nanorobots able to self-copy and replicate. Potentially able to swim around in one's bloodstream these nanorobots will do countless numbers of cellular repairs. Medicine will undergo a drastic change, by maintaining health at the cellular level. Nanotechnology experts contend these nanorobots could extend human longevity to hundreds of years by simply keeping cells in the human body replicating in normal ways. Thus, no more aging! Disease will no longer concern the social actor as the nanorobot can travel to the site of tissue damage or disease, and eradicate problems while regenerating cellular growth to build new healthy tissue or body parts.

Nano-manufacturing will be available from common elements already in your own backyard, at little or no cost. Nano-assemblers will move atoms and nanorobots at the atomic level and allow for the creation of any commodity you can possibly dream about. We will be able to manufacture "steak" just like cattle, utilizing the atomic components of oxygen, carbon, and grass, having the nanorobots couple the atoms together in the exact same order as cattle do the natural way to make steak via nanotechnology. Food could be replicated. Starvation and hunger could be eliminated from the globe. Homelessness can disappear: programme the nano-assemblers and watch them construct a mansion before your very eyes, using atoms like those found in wood linking them in the correct atomic sequences. Manufacturing comprised from simple soil in one's own backyard, would yield almost any commodity. This means raw material extraction, currently necessary and contracted globally, would not be required. Most needed atoms and raw material can be extracted with a wheel barrow and a shovel at home in your backyard. This could potentially change the shape of all global manufacturing and trade.

Nanotechnology could also clean up the environment by eliminating polluting atoms or molecules that are given off in chemical reactions. Nanorobots will be able to "clean-up" any toxic site or oil spill. This will occur on an atomic level which will prevent harm to all ecosystems. Sustainability, now only a pious hope, could become a reality; or, to put it another way, it can usher in The Third Revolution proposed by Paul Harrison. Harrison argues that global destruction is just moments away, and uses the metaphor of Hamlet with only 30 min to live before he dies or acts. Without a third revolution, Harrison suggest, the planet Earth will die from pollution! Nanotechnology offers the badly needed solutions that Harrison demands as it could guarantee sustainability and put an end to food shortages and pollution simply by letting the Nan robots do their work. Drexler sees this as a very realistic proposition.

Computers, medicine, machinery, economics, the environment, and many facets of society will feel the shock waves of this technological nanorevolution. A number of interesting questions arise. How will economics, as we know it, be transformed? How will the consumer society fundamentally change when influenced by nanotechnology as a catalyst? Could one imagine a future society, with the potential to construct and manufacture anything one wishes? Inequality, poverty, and social class might be re-defined if a shift in jobs and commodification evolves! If social change occurs due to nanotechnology, how will that alter the way society is constructed today? What new theory from a sociological perspective will be needed? How would older accepted social theory be reworked? For instance, if nanotechnology was to eliminate poverty, Marxist Social Conflict Theory would not explain society very accurately, so what new theory might emerge that better explains shifts in social structure? In analyzing Social Conflict Theory if the concept of social class is identified not as an inequality based on economic determinism, but as equal access to social reward due to nanotechnology, the classless society may suddenly evolve past the vision of Marxism requiring revision.

Structural Functionalism Theory and Symbolic Interaction Theory are theoretical sets of explanations of society and relationships that would also require reconfiguration. The work of Talcott Parsons or George H. Mead could be revisited due to their potential lack of clarity when explaining a new social order. Old theory will fall by the wayside and new theorists are certain to emerge. Social actors will have new roles. How will their roles be reconstructed? Will family patterns change if work and education changes? How will religion evolve to meet the moral crisis that might emerge? What jobs become important, and which fields will take on added importance, which ones will be promoted, and which ones will receive the ever scarce funding resources.

These future nanotechnology changes will establish new social roles according to new social structures of society. Changes in employment patterns, will lead to role changes:

credentialism would change if nanotechnology changes the workplace. Demand for study in particular fields would shift from one discipline to another due to job growth in different work sectors, due to nanotechnology. This holds the potential to reshape intellectual life, since college enrollments will change due to new type of jobs.

We may seek answers of a spiritual nature as well as a social and scientific nature. Future "Creation of Humans" from nanotechnology might one day be possible leading society to question concepts of God and morality. Will these so-called perfect beings created by nanorobots have "souls" or not? Cloning has recently brought this moral question to the fore; yet with nanotechnology cloning will become an outdated technology that takes too much time to produce body parts or a person when nanorobots can do it significantly quicker. The cloning process shown in films like Gattica is already out of date. Nanotechnology will allow full scale cloning without the need of genetic engineering.

We can predict, first, that nanotech innovations will raise serious ethical issues in the present and upcoming decades, and we should be thankful to the commentators for beginning discussion. The unique properties of many nanoparticles and materials that scientists are beginning to exploit raise significant health and environmental concerns. Second, advances in nanomedicine will probably be accompanied by ethical concerns about human enhancement, rising health care costs and rationing, and increasing diagnostic powers. Third, nano-enhanced tracking devices may threaten individual privacy, especially if they become ubiquitous. Finally, nanotechnology will likely raise justice-related concerns about the economic effects of a technological revolution.

But we must resist equating new technological powers with novel ethical challenges. These ethical issues are already raised by other technologies, and we would waste resources and forget lessons already learned by unreflectively assuming that nanotechnology requires us to invent a whole new ethics—as if that were possible—with its attendant conferences, journals, centres, and funding mechanisms. Moreover, even if the most radical futuristic visions of nanotechnology are realised, history teaches us that such extreme prophecies should not, at this time, flame debate or warrant ethical attention.

There is no settled definition of "nanotechnology," but all the candidates say something about understanding and manipulating matter on the scale of one to one hundred nanometers. However, that criterion alone subsumes much of conventional science and technology, including most chemical reactions. If nanotechnology represents a new field, it should refer to the exploitation of newly discovered laws and properties that nanoparticles uniquely possess, such as their conductivity and quantum effects.

Even without a precise definition, nanotechnology holds great promise for human welfare both on its own and through its convergence with other technologies. Nanomedical research already shows signs of that potential. Cancer researchers have successfully tested in mice a nanocell drug delivery system that enables a slow release of antiangiogenesis agents and chemotherapy. DNA nanoparticles have been used to accomplish gene transfer and improve physiological functioning in patient-subjects with cystic fibrosis. Researchers have nanoengineered membrane systems designed to clean water efficiently by targeting pollutants. Other nanotechnologies are likely to produce stronger, lighter materials and faster, more efficient computers.

Despite the potential of manipulating matter at the nanoscale, Eric Drexler, nanotechnology's best-known visionary, laments its association with that broadly construed endeavor. On his view, we are losing sight of nanotechnology's truly radical promise—molecular manufacturing—first imagined by Richard Feynman forty-five years ago: "It would be, in principle, possible for a physicist to synthesize any chemical substance that the chemist writes down. Put the atoms down where the chemist says, and so you make the substance." The essence of nanotechnology, on Drexler's view, lies in the use of self-replicating assemblers that can carry out Feynman's vision by guiding chemical reactions that result in any desired atomic configuration. Disassemblers can also break down rock and other raw materials, which will then be reshaped into anything we want. In Drexler's words, we could "grow spaceships from soil, air, and sunlight."

Initially, the ethical concerns about nanotechnology focused on Drexler's vision. Commentators expressed fantastic fears of self-replicating nanobots breaking down all matter, turning the world into "gray goo." Recent dialogue has centered on more realistic risks. But with the primary focus on risks, some activist groups call for a moratorium on research involving nanoparticles until we understand their hazards, and some commentators go even further, urging nanotech research to cease completely.

Moratorium advocates have raised public awareness about health and safety concerns, but their goal is unrealistic. Industrial nations have invested billions in nanotechnology research and development. Further, any attempt to define the boundaries of a moratorium will be impossible. Nanotechnology will either be defined too narrowly to prevent circumvention, or too broadly, prohibiting conventional research.

The difficulty in drawing a bright-line definition represents a problem for any generalized approach to nanotech ethics. But ethics should not be concerned with whether a certain technological project falls squarely within an accepted definition of "nanotechnology." There may be scientific reasons to view some research as "nanotechnology," but what matters ethically are the potential harms and benefits to human welfare and the environment. Ethical attention must be directed at specific risks of particular innovations; it should not try to address nanotechnology as a whole.

A single framework specifically created to analyse all nanotech-related ethical issues and risks is impossible for reasons other than definitional difficulties. A unified framework would need to specify interests that may be threatened and benefited by nanotechnology. But nanotech innovations will emerge in diverse fields where different interests and values are implicated. It would not be helpful or important to analyse under a unified framework the risks of developing nano-enhanced weaponry and the risks nanotechnology may pose to certain manufacturing industries and workers. Academics, policy-makers, and practitioners in different fields apply ethical principles and conduct risk analyses regularly, and there is no reason to alter them merely because the technology's size will introduce new facts for consideration.

An issue-by-issue ethical approach is not only required, but is sobering. None of the ethical concerns associated with nanotechnology is unprecedented, and none raises novel ethical issues or demands new ethical principles. Resolving these concerns will require weighing and balancing reasons related to the values we have long embraced—autonomy, beneficence, fairness, efficiency, and environmental preservation.

The most pressing concern is whether nanomaterials pose dangers to health and the environment. The qualitatively different behaviour of nanoparticles accounts for both nanotech's promise and its risks. The ability of some nanoparticles to cross the blood-brain barrier will benefit drug delivery but may present serious workplace dangers. We must determine whether current regulations, which consider only the macroscale behaviour of chemical substances, should be amended to accommodate their nanoscale properties. An acceptable level of workplace exposure to fullerenes may differ from safe levels for other forms of carbon.

But the need for toxicology and environmental risk analyses did not arise with nanotechnology: people have been conducting experiments to discern the toxicity of substances for centuries, and these experiments have always required an ability to take on new insights into the natural world. Modern toxicologists study radiation, microbes, organophosphates, new pharmaceuticals, and other substances, discerning how they are absorbed, metabolized, and distributed through bodies and determining what the appropriate exposure limit is. Elucidating safety concerns about nanomaterials and particles does not require "nanoethicists" to offer novel ethical arguments, but rigorous scientific research to identify the effects of these substances on biological systems, as well as safe ways to work with and transport them. The possibility that the effects of nanomaterials will only become visible years from now is hardly unique to them. It accompanies technological progress, which we can and must constantly monitor.

Nanomedical issues do not require novel ethical theorizing, either. The convergence of nanotech, bioetech, information technology, and cognitive science may empower us to enhance our cognitive abilities significantly, and ethical debate about the desirability of human enhancement is now already well under way, with the work of the President's Council on Bioethics representing the most widely known discussion. Whether enhancement is based on biotechnology or its convergence with nano- and other technologies, the relevant values and moral principles are the same.

Nanotechnology may improve our medical diagnostic capabilities, which could raise difficult ethical issues; but these issues are familiar because of extensive ethical research on genetics. We grapple, with the question of what genetic knowledge to give patients and research subjects, especially when no cures exist for the relevant diseases, and with the insurance-related danger of being able to predict an individual's future health. The values and moral principles that are relevant to these dilemmas will not change when the underlying technology is nano.

Finally, as our experience with technological progress demonstrates, we can predict that nanomedical advances are likely to increase health care costs and create rationing dilemmas. Therapeutic and diagnostic measures developed for narrow causes, in which the benefits clearly outweigh the costs, will then be applied more broadly when the cost-benefit ratio is at best marginal. But to repeat the refrain, whether rising costs and rationing dilemmas are caused by current technology or nanomedical innovations, we will address them by appeal to the same values and ethical principles.

Potential privacy threats are not only common to other technologies, but primarily require strategic policies—not new ethical principles—to achieve ends about which there is no ethical disagreement. Individual privacy is an oft-cited nanotech concern because the technology may produce cheap, invisible devices that could be implanted in clothing and household appliances to gather and wirelessly transmit information. Private health data would be of particularly serious concern, especially if diagnostic nanodevices could be embedded directly in human bodies. But the Internet—where users' personal information is collected via e-commerce transactions, cookies, spyware, and so on—presents similar problems. The important thing is to find ways of preventing unwanted invasion of privacy and abuse of personal information. Privacy-enhancing technologies, such as encryption tools and cookie cutters, have developed along with the Internet, and they must now keep pace with whatever threats nanotechnology poses. By the time we face these threats, we may be able to draw on our experience with information technology to think about whether industry self-regulation is preferable to government regulation and what kind of legal protections are most effective.

A similar analysis is appropriate for justice-related concerns, including the possibility of a "nanodivide," on one side of which the wealthy nations reap nanotechnology's economic, medical, and other benefits, while the poorest, on the other side, fall ever farther behind. Other technologies have also increased global inequality. We need not reflect upon a specifically "nano" divide to unveil a duty of the wealthiest to alleviate the poverty and suffering of the poorest. What will be needed is what we need now, with the technology we have: policies that ensure that the poorest benefit from advances in generating cheap energy, filtering water, curing disease, and producing food.

If the most radical futuristic visions prove possible, and if we allowed them to be realised, then new ethical issues might certainly arise. Societal transformation would be absolutely radical if each household had its own Drexlerian manufacturing kit, with self-replicating nanobots turning carbon and sunlight into whatever matter was needed, including food. Such abundance would render current theories of justice irrelevant, since they all assume a scarcity of resources. The very structure of our economy would be drastically transformed if virtually any good could be produced without labour.

However, we should not spend resources developing the ethics for a Drexlerian world. Predictions about the underlying science and technology are simply too speculative. The history of futurism is fraught with fantastic mistakes by great minds. John von Neumann foresaw global warming, but predicted that by now, nuclear energy would "be free—just like the unmetered air." Many of the predictions made in 1975 by Asilomar conferees about recombinant DNA either have never been realised or took longer than expected, and very few of the scientists foresaw the technology's actual or positive ramifications.

Predicting nanotechnology's long-term future is impossible because it requires foreseeing how it will affect society and how, in return, societal and economic forces will shape it. But just taking nanoscience in isolation, we do not know whether the radical control over nature implied by molecular manufacturing is possible. To break and make atomic bonds on the massive scale required for manufacturing consumer goods, we would need to harness enormous amounts of energy efficiently and cheaply. Nanotechnology, converging with photovoltaics, may make it possible, but it is too speculative. If we are discovering a new physics, then decisive claims about what is possible or impossible may be premature. But acknowledging that these futuristic visions have some plausibility does not warrant resources for a new discipline of nanoethics. Even if they are plausible, new, nano-tailored values will not be needed. And anyway, there are serious problems now and on a more probable horizon.

Some commentators say nanotech ethics has an urgent need to "catch up" to the science: "Unfortunately, we are just beginning to address these issues in a nuanced way.... To think through the challenges ahead, we will need the same kind of exponential growth in ethics research that is taking place in nanotechnology." In fact, the most frequently cited ethical concerns are already under consideration because of other technologies, and though they will continue to warrant attention, nuanced ethical reflection cannot begin on the new twists that nanotechnological developments will introduce until we see the details.

Extensive safety and environmental studies are what's most crucial right now. Six federal agencies are leading those efforts under the National Nanotechnology Initiative, which aims to facilitate communication of toxicology and environmental findings and to assess continually the research needed to adapt regulations to nanoparticles' characteristics. Because assessing risk involves value judgments as well as empirical data, the National Institute of Standards and Technology is developing relevant standards. Our primary ethical challenge will be gauging the adequacy of the toxicology studies and the developed standards for introducing nanomaterials into common usage.

Preparing for the unknown is also necessary. The National Nanotechnology Initiative's stated commitment to continual assessment of nanotechnology's societal implications is important, given history's lessons. The world is unlikely to become gray goo, but there will be unforeseeable consequences. There are with any technology. We can expect some kinds of unintended, cumulative, chronic problems, analogous to the environmental legacies of nuclear energy and chlorofluorocarbons. Specific materials and manufacturing innovations may upend economies and ruin livelihoods, and developments in medicine, forensics, and weaponry will affect medical, legal, and political systems. We cannot detail in advance the transformative effects of nanotechnology, but closely monitoring its developments may help us reap its benefits while minimizing its harms.

Finally, the scientific community should prepare for a public dialogue. Public debate has the potential to gravitate toward the poles of the most optimistic futurists and the most pessimistic luddites. Scientists have a stake in preventing unjustified fears from prevailing, but they also have a duty to prevent public dialogue from turning into an "apocalyptic nano-drama." As one nano-commentator points out, that kind of debate will preclude serious discussion by obscuring the complexity of real ethical problems and fading them into the background. For ethical debate to serve its purpose, it must be properly informed by the science.

Advances in nanomedicine offer the possibility of new and intriguing opportunities in NP-based early detection, diagnosis, and treatment of diseases. Commercial development of nanotechnology and its many applications may unleash a spectrum of human health hazards that at the present time can only be speculated about without any detailed understanding of the toxic nature of NPs. The review of literature in the fields of toxicology and the possible human health effects of UFPs and NPs provide only a glimpse of some toxic paradigms that compel us to weigh the adverse effects against the beneficial effects. Because we know little about the toxic health hazards of NPs in vivo and in vitro, pharmacokinetic and toxicologic studies are mandatory before large-scale industrial production and use are implemented. In this regard the U.S. Environmental Protection Agency, the International Life Sciences Institute Research Foundation, and the Risk Science Institute convened working groups comprising experts in the fields of nanotechnology from academia and government to develop new toxicity screening, reporting, and hazard identification of engineered nanomaterials. Although at this time, the benefits of nanotechnology dominate our thinking, the potential for undesirable human health outcomes should not be overlooked. Consistently large numbers of studies have reported associations between UFP exposure and morbidity in elderly and compromised individuals. Furthermore, recent studies also emphasize the impact of day-to-day variations in particle concentrations and exposures for short periods as important factors in cardiac events in predisposed population. Therefore, there is reason to suspect that

NPs with size and surface characteristics similar to UFPs are likely to cause diseases—some with a long latency. With widespread industrialization of nanotechnology, there is the potential for ambient air pollution and a conceivable threat to the general population.

Doubtless, nanotechnology will have a profound impact on a wide range of applications and therefore on many aspects of human life, including environmental decontamination, water purification, cheaper electricity, and better disease treatment modalities. One major challenge facing industry and government is the lack of information on the possible adverse health effects caused by exposure to different nanomaterials. Development of safety guidelines by government for the nanotechnology industries, including manufacturing, monitoring of worker exposure, ambient release of NPs, and risk evaluations, is mandatory to promote nanotechnology for its economic incentives and medicinal applications.

GERM THEORY AND ANTISEPSIS

By the middle of the 19th century, pain had been banished from surgical operations, but one grave danger still faced every patient submitting himself to the surgeon's knife. This was the ever-present risk of sepsis. Hospital diseases such as erysipelas, pyemia, septicemia and gangrene, were rife. In the 1850s the death rate after amputations varied from 25 Percent-60 Percent in different countries and in military practice it reached the appalling figure of 75 Percent-90 Percent. The first ovariotomies, which were the first abdominal operations performed on a fairly large scale, had a mortality rate of more than 30 Percent even in the most expert hands. That all these diseases were due to some form of "contagion" had long been suspected, but the general view was that whatever agent was responsible was generated spontaneously in wounds. Alternatively, it was theorized that air itself was responsible for suppuration and many attempts were made to exclude the air from wounds by means of elaborate dressings.

Some medical men had postulated the existence of minute particles in the air which carried contagion, the so-called "germ theory." Girolamo Fracastoro of Verona proposed "seminaria, the seeds of disease which multiply rapidly and propagate their life," minute bodies passing unseen from the infector to the infected by contact, by clothing or utensils, and by infection at a distance through the air.

What made the germ theory of contagion so difficult to accept was that no one could see the supposed microbes. Magnifying lenses were used in ancient times, and by the beginning of the 17th century they had been combined in a tube to make the compound microscope. The first man to employ the microscope in investigating the causes of diseases was probably Athanasius Kircher, a learned Jesuit priest. In 1658 Kircher described experiments upon the nature of putrefaction, showing how maggots and other living creatures developed in decaying matter. He also claimed to have found in the blood of plague-stricken patients "countless masses of small worms, invisible to the naked eye." It is impossible that he could have seen the plague bacillus with the very low power microscopes at his disposal, but he may have seen some of the larger microorganisms and his statements about the doctrine of contagion are even more explicit than those of Fracastoro.

The great pioneer of modern microscopy was Anthoni van Leeuwenhoek, a Dutch linen draper, who ground his own lenses and made hundreds of microscopes. Few of his instruments provided a magnification of more than 160X but he is generally credited as making the first observations of germs, reported in his communications to the Royal Society in London. Leeuwenhoek was the first to describe spermatozoa and he gave the first complete account of the red blood corpuscles in 1674; he also found that the film from his own teeth contained "little animals, more numerous than all the people in the Netherlands."

Ignaz Philipp Semmelweis, an assistant at the Vienna General Hospital in the maternity clinic, was investigating the 29 Percent postpartum mortality rate among women in Ward

One, where births were handled by medical students, vs. a 3 Percent rate in Ward Two where births were handled by midwifery pupils. Semmelweis noticed that the appearances of these deaths from puerperal fever looked the same as those observed in the body of an older colleague, forensic medicine Professor Jakob Kolletschka, who had died from adissection wound suffered in the clinic. He correctly surmised that the postpartum deaths were caused by infection from "putrid particles" carried on the hands of medical students who often shuttled back and forth between the labour wards, the obstetrical clinic, and the autopsy room where dissections were performed. Semmelweis instituted a simple routine of hand-washing with water solutions of chloride of lime, which promptly reduced mortality to 1.27 Percent. Unfortunately, his ideas were met with fierce opposition by the conservative medical community in Vienna, who subjected him to laughter and ridicule. In disgust, Semmelweis left Austria for Budapest. There he became head of the obstetrical division of St. Rochus Hospital, where puerperal fever mortality rates were reduced to below 1 Percent after Semmelweis introduced chlorinated water disinfection.

The man who elucidated the true nature of infection, founded the science of bacteriology, and paved the way for Lister and the antiseptic system in surgery was Louis Pasteur. Pasteur was led to his great discoveries regarding bacteria and other microorganisms by his investigations into the process of fermentation. He showed conclusively that fermentation was brought about by some external agent entering the wine. He proved by painstaking experiments under rigorously controlled conditions that meat and fluids like blood did not putrefy if they were kept in such a way that all air was excluded from them. By taking samples of air at different levels Pasteur showed that contamination became less with increasing altitude. Then he proved that the contaminating agents were living organisms which were everywhere – in every room, in the air, on every article of clothing, on furniture, on the ground, and on the skin. He showed that putrefaction was caused by the presence of bacteria and that this applied to putrefaction in foods, urine, and in wounds.

The application of Pasteur's discoveries to surgical practice was the work of Joseph Lister, a young English surgeon who had concluded that it must also be bacteria that caused the suppuration, pus and gangrene which plagued the surgical wards of those days. He determined to prevent the access of organisms by killing them in or on the surface of the wound. Pasteur had shown that heat could kill microbes, but it was impossible to apply heat to a wound without burning the patient, so some chemical substance had to be found. After trying various chemical agents he finally selected carbolic acid, and he insisted that everything which touched the wound, the dressings, the instruments and the fingers, should be treated with this antiseptic. He even produced an antiseptic atmosphere by means of a carbolic spray. The clinical results of Lister's first antiseptic system in 1865 included 11 compound fracture cases with only one death, a 9 Percent mortality rate, marking a watershed between the primitive and modern eras of surgery.

Throughout the 19th century, the arguments continued as to whether microorganisms seen in a sick patient were merely coincidental with the illness, or resulted from the changes brought about by the illness itself. In 1882 the German microbiologist Robert Koch formulated three famous postulates to guide scientists searching for disease-causing microbes. Koch argued that to prove an organism causes a disease, microbiologists must show that the organism occurs in every case of the disease; that it is never found as a harmless parasite associated with another disease; and that once the organism is isolated from the body and grown in laboratory culture, it can be introduced into a new host and produce the disease again. Koch and his pupils discovered specific bacillary causes for various diseases, including anthrax, cholera, tuberculosis, gonorrhea, diphtheria, leprosy, typhoid, trypanosomiasis, and malaria.

The theory that specific germs could cause specific disease remained contentious until the beginning of the 20th century. Many scientists of great repute rejected Koch's conclusions, with one scientist confidently asserting that "no microbe found in the living blood of any animal was pathogenic." In one celebrated case, Max von Pettenkofer of Bavaria, a distinguished 19th century experimental hygienist, induced Koch to send him a sample of his cholera vibrios culture and then wrote a letter back to Koch in 1892, as follows:

"Herr Doctor Pettenkofer presents his compliments to Herr Doctor Professor Koch and thanks him for the flask containing the so-called cholera vibrios, which he was kind enough to send. Herr Doctor Pettenkofer has now drunk the entire contents and is happy to be able to inform Herr Doctor Professor Koch that he remains in his usual good health."

Apparently Pettenkofer, aged 74 at the time he wrote the letter, survived this cholera exposure quite well, perhaps possessing the high stomach acidity which sometimes neutralizes the bacillus, though he shot himself to death in Munich 9 years later.

Skeptics notwithstanding, microbes were key. Enormous new vistas now lay open, as surgeons could confidently make an incision through intact skin without incurring an extreme risk of wound infection. The next step was to progress beyond killing wound bacteria with chemical antiseptics, to the prevention of bacterial contamination by eliminating bacteria in the operating theater – aseptic surgery. The use of steam sterilization of instruments, dressings and gowns, the wearing of masks, caps and gloves, air filtration and the other rituals of the operating theater of today were introduced over the decades following Lister's efforts.

Anesthesia and antisepsis enabled surgeons to carry out procedures that had formerly been quite beyond them, including long operations inside the head, the abdomen and the pelvis. Theodor Billroth of Vienna resected the esophagus in 1872, parts of the intestines in 1878, and the pyloric end of the stomach in 1881. Billroth also made the first complete excision of the larynx. The first successful repair of a gunshot wound on a major artery was performed in Chicago by John B. Murphy in 1897; Ludwig Rehn of Frankfurt am Main performed the first successful repair of a cardiac injury in Germany in the same year.

Writing in 1874, Sir John Eric Erichsen, Professor of Surgery at University College, London, had predicted that "the abdomen, the chest, and the brain would be forever shut from the intrusions of the wise and humane surgeon." By the time of Erichsen's death 22 years later, surgeons had successfully removed from patients the stomach and large parts of the intestines, a whole lung had been excised, and a brain tumor had been extirpated. These operations were not mere feats of surgical showmanship; they saved the lives and restored the health of thousands of human beings.

BLOOD TRANSFUSIONS

Vague references to blood transfusion are found in medieval writings, though physicians historically were far more concerned with taking blood out of the body than with putting it back in. One story tells of an attempt to prolong the life of Pope Innocent VIII by means of a blood transfusion. By one account, a Jewish physician transfused the aged Pontiff with blood from three small boys, who each received one ducat as their reward; but another account says the blood was drunk, not infused. One of the earliest proposals to transfuse blood was by Andreas Libavius, a physician of Halle in Saxony, in 1615.

The first serious attempts at blood transfusion were made in England and France. Sir Christopher Wren carried out numerous experiments on the injection of liquids into the veins of animals. Richard Lower of Oxford carried out the first successful transfusion from artery to vein between two dogs in 1665, a feat repeated by the French physician Jean-Baptiste Denys, who went on to attempt transfusion of blood from one kind of animal into another. Finally, before the Royal Society in 1667, Lower transfused a human youth, 15 years of age, from a sheep. The patient was greatly improved and the only ill effect was a feeling of great heat along his arm; subsequent patients were not so lucky. Blood transfusions from lambs and calves to humans continued to be tried in the mid 17th century, but were not very successful, and in 1670 were finally forbidden by law in England.

The first known transfusion of human blood into an already moribund person was attempted by James Blundell in 1818, and on several other occasions during the 1820s, with poor results. Transfusion was carried out on a small scale during the American Civil War, but technical difficulties connected with premature clotting and the occurrence of accidents arising from the use of incompatible blood prevented the rapid acceptance of the process. The cause of many of the untoward effects of blood transfusion was finally explained in 1901 when the presence of agglutinins and iso-agglutinins in the blood was demonstrated by Karl Landsteiner in Vienna, winning him the Nobel Prize in 1930; in 1907, the four main blood groups were determined by Jan Jansky of Prague. Rhesus factor was later identified by two American scientists, Philip Levine and Rufus Stetson. These advances were of fundamental importance and it became possible in the 20th century to eliminate most of the fatalities due to incompatibility, allowing blood transfusions to be practiced reliably and with uniformly good results.

21ST CENTURY MEDICINE

It is always somewhat presumptuous to attempt to predict the future, but in this case we are on solid ground because most of the prerequisite historical processes are already in motion and all of them appear to be clearly pointing in the same direction.

Medical historian Roy Porter notes that the 19th century saw the establishment of what we think of as scientific medicine. From about the middle of that century the textbooks and the attitudes they reveal are recognizable as not being very different from modern ones. Before that, medical books were clearly written to address a different mind-set.

But human health is fundamentally biological, and biology is fundamentally molecular. As a result, throughout the 20th century scientific medicine began its transformation from a merely rational basis to a fully molecular basis. First, antibiotics that interfered with pathogens at the molecular level were introduced. Next, the ongoing revolutions in genomics, proteomics and bioinformatics provided detailed and precise knowledge of the workings of the human body at the molecular level. Our understanding of life advanced from organs, to tissues, to cells, and finally to molecules, in the 20th century. By the early 21st century, the entire human genome will be mapped. This map will inferentially incorporate a complete catalog of all human proteins, lipids, carbohydrates, nucleoproteins and other molecules, including full sequence, structure, and much functional information. Only some systemic functional knowledge, particularly neurological, may still be lacking by that time.

This deep molecular familiarity with the human body, along with simultaneous nanotechnological engineering advances will set the stage for a shift from today's molecular scientific medicine in which fundamental new discoveries are constantly being made, to a molecular technologic medicine in which the molecular basis of life, by then well-known, is manipulated to produce specific desired results. The comprehensive knowledge of human molecular structure so painstakingly acquired during the 20th and early 21st centuries will be used in the 21st century to design medically-active microscopic machines. These machines, rather than being tasked primarily with voyages of pure discovery, will instead most often be sent on missions of cellular inspection, repair, and reconstruction. In the coming century, the principal focus will shift from medical science to medical engineering. Nanomedicine will involve designing and building a vast proliferation of incredibly efficacious molecular devices, and then deploying these devices in patients to establish and maintain a continuous state of human healthiness.

The very earliest nanotechnology-based biomedical systems may be used to help resolve many difficult scientific questions that remain. They may also be employed to assist in the brute-force analysis of the most difficult three-dimensional structures among the 100,000-odd proteins of which the human body is comprised, or to help ascertain the precise function of each such protein. But much of this effort should be complete within the next 20-30 years because the reference human body has a finite parts list, and these parts are already being sequenced, geometered and archived at an ever-increasing pace. Once these parts are known, then the reference human being as a biological system is at least physically specified to completeness at the molecular level. Thereafter, nanotechnology-based discovery will consist principally of examining a particular sick or injured patient to determine how he or she deviates from molecular reference structures, with the physician then interpreting these deviations in light of their possible contribution to, or detraction from, the general health and the explicit preferences of the patient.

In brief, nanomedicine will employ molecular machine systems to address medical problems, and will use molecular knowledge to maintain human health at the molecular scale.

VOLITIONAL NORMATIVE MODEL OF DISEASE

What exactly is "medicine"? Dictionaries give several definitions, ranging from the very restrictive to the most general, as follows: "a drug or remedy" "any substance used for treating disease" "any drug or other substance used in treating disease, healing, or relieving pain" "in a restricted sense, that branch of the healing art dealing with internal diseases" "treatment of disease by medical, as distinguished from surgical, treatment" "the branch of this science and art that makes use of drugs, diet, etc., as distinguished especially from surgery and obstetrics"; "the study and treatment of general diseases or those affecting the internal parts of the body"; "the science of treating disease, the healing art" "the art and science of preventing or curing disease"; "the act of maintenance of health, and prevention and treatment of disease and illness"; "the department of knowledge and practice dealing with disease and its treatment"; or, most generally, "the science and art of diagnosing, treating, curing, and preventing disease, relieving pain, and improving and preserving health".

The contemporary physician might at first be inclined to relegate molecular approaches to some minor subfield, perhaps "nanoanalytics," "nanogenomics," or "nanotherapeutics." This would be a serious mistake, because the application of molecular approaches to health care will significantly impact virtually every category of laboratory and clinical practice across the board. Thus we are led to the broadest possible conception of nanomedicine as "the science and technology of diagnosing, treating, and preventing disease and traumatic injury, of relieving pain, and of preserving and improving human health, using molecular tools and molecular knowledge of the human body."

This brings us to the question of "disease," a complex term whose meaning is still hotly debated among medical academics. The results of a survey of four different groups of people who were read a list of common diagnostic terms and then asked if they would rate the condition as a disease. Illnesses due to microorganisms, or conditions in which the doctor's contribution to the diagnosis was important, were most likely to be called a disease, but if the cause was a known physical or chemical agent the condition was less likely to be regarded as disease; general practitioners also had the broadest definition of disease.

No less than eight different types of disease concepts are held by at least some people currently engaging in clinical reasoning and practice, including:

1. *Disease Nominalism*: A disease is whatever physicians say is a disease. This approach avoids understanding and forestalls inquiry, rather than furthering it.
2. *Disease Relativism*: A disease is identified or labeled in accordance with explicit or implicit social norms and values at a particular time. In 19th century Japan, armpit odor was considered a disease and its treatment constituted a medical specialty. Similarly, 19th-century Western culture regarded masturbation as a disease, and in the 18th century, some conveniently identified a disease called drapetomania, the"abnormally strong and irrational desire of a slave to be free." Various non-Western cultures having

widespread parasitic infection may consider the lack of infection to be abnormal, thus not regarding those who are infected as suffering from disease.

3. *Sociocultural Disease*: Societies may possess a concept ofdisease that differs from the concepts of other societies, but the concept may also differ from that held by medical practitioners within the society itself. For instance, hypercholesterolemia is regarded as a disease condition by doctors but not by the lay public; medical treatment may be justified, but persons with hypercholesterolemia may not seek treatment, even when told of the condition. Conversely, there may be sociocultural pressure to recognize a particular condition as a disease requiring treatment, such as alcoholism and gambling.
4. *Statistical Disease*: A condition is a disease when it is abnormal, where abnormal is defined as a specific deviation from a statistically-defined norm. This approach has many flaws. A statistical concept makes it impossible to regard an entire population as having a disease. Thus tooth decay, which is virtually universal in humans, is not abnormal; those lacking it are abnormal, thus are "diseased" by this definition. More reasonably, a future highly-aseptic society might regard bacterium-infested 20th century humans who contain in their bodies more foreign microbes than native cells; as massively infected. Another flaw is that many statistical measurables such as body temperature and blood pressure are continuous variables with bell-shaped distributions, so cutoff thresholds between "normal" and "abnormal" seem highly arbitrary.
5. *Infectious Agency*: Disease is caused by a microbial infectious agent. Besides excluding systemic failures of bodily systems, this view is unsatisfactory because the same agent can produce very different illnesses. Infection with hemolytic *Streptococcus* can produce

diseases as different as erysipelas and puerperal fever, and Epstein-Barr virus is implicated in diseases as varied as Burkitt's lymphoma, glandular fever, and naso-pharyngeal carcinoma.

6. *Disease Realism*: Diseases have a real, substantial existence regardless of social norms and values, and exist independent of whether they are discovered, named, recognized, classified, or diagnosed. Diseases are not inventions and may be identified with the operations of biological systems, providing a reductionistic account of diseases in terms of system components and subprocesses, even down to the molecular level. One major problem with this view is that theories may change over time — almost every 19th century scientific theory was either rejected or highly modified in the 20th century. If the identification of disease is connected with theories, then a change in theories may alter what is viewed as a disease. The 19th century obsession with constipation was reflected in the disease labelled "autointoxication," in which the contents of the large bowel were believed to poison the body. Consequently much unnecessary attention was paid to laxatives and purgatives and, when surgery of the abdomen became possible toward the end of the century, operations to remove the colon became fashionable in both England and America.

7. *Disease Idealism*: Disease is the lack of health, where health is characterized as the optimum functioning of biological systems. Every real system inevitably falls short of the optimum in its actual functioning. But by comparing large numbers of systems, we can formulate standards that a particular system ought to satisfy, in order to be the best of its kind. Thus "health" becomes a kind of Platonic ideal that real organisms approximate, and everyone is a less than perfect physical specimen. Since we are all flawed to some extent, disease is a matter of degree, a more or

less extreme variation from the normative ideal of perfect functioning. This could be combined with the statistical approach, thus characterizing disease as a statistical variation from the ideal. But this view, like the statistical, suffers from arbitrary thresholds that must be drawn to qualify a measurable function as representing a diseased condition.

8. *Functional Failure*: Organisms and the cells that constitute them are complex organized systems that display phenomena resulting from acting upon a programme of information. Programmes acquired and developed during evolution, encoded in DNA, control the processes of the system. Through biomedical research, we write out the programme of a process as an explicit set of instructions. There are completely self-contained "closed" genetic programmes, and there are "open" genetic programmes that require an interaction between the programmed system and the environment, e.g. learning or conditioning. Normal functioning is thus the operation of biologically programmed processes, e.g. natural functioning, and disease may be characterized as the failure of normal functioning. One difficulty with this view is that it enshrines the natural as the benchmark of health, but it is difficult to regard as diseased a natural brunette who has dyed her hair blonde in contravention of the natural programme, and it is quite reasonable to regard the mere possession of an appendix as a disease condition, even though the natural programme operates so as to perpetuate this troublesome organ.* A second weakness of this view is that disease is still defined against population norms of functionality, ignoring individual differences. As a perhaps overly simplistic example, 65 Percent of all patients employ a cisterna chyli in their lower thoracic lymph duct, while 35 Percent have no cisterna chyli – which group has a healthy natural programme, and which group is "diseased"?

The author proposes a ninth view of disease, a new alternative which seems most suitable for the nanomedical paradigm, called the "volitional normative" model of disease. As in the "disease idealism" view, the volitional normative model accepts the premise that health is the optimal functioning of biological systems. Like the "functional failure" view, the volitional normative model assumes that optimal functioning involves the operation of biologically programmed processes.

However, two important distinctions from these previous views must be made. First, in the volitional normative model, normal functioning is defined as the optimal operation of biologically programmed processes as reflected in the patient's own individual genetic instructions, rather than of those processes which might be reflected in a generalized population average or "Platonic ideal" of such instructions; the relative function of other members of the human population is no longer determinative. Second, physical condition is regarded as a volitional state, in which the patient's desires are a crucial element in the definition of health. This is a continuation of the current trend in which patients frequently see themselves as active partners in their own care.

In the volitional normative model, disease is characterized not just as the failure of "optimal" functioning, but rather as the failure of either (a) "optimal" functioning or (b) "desired" functioning. Thus disease may result from:

1. A failure to correctly specify desired bodily function (specification error by the patient),
2. A flawed biological programme design that doesn't meet the specifications (programming design error),
3. Flawed execution of the biological programme (execution error),
4. External interference by disease agents with the design or execution of the biological programme (exogenous error), or
5. Traumatic injury or accident (structural failure).

In the early years of nanomedicine, volitional physical states will customarily reflect "default" values which may differ only insignificantly from the patient's original or natural biological programming. With a more mature nanomedicine, the patient may gain the ability to substitute alternative natural programmes for many of his original natural programmes. The genes responsible for appendix morphology or for sickle cell expression might be replaced with genes that encode other phenotypes, such as the phenotype of an appendix-free cecum or a phenotype for statistically typical human erythrocytes. Many persons will go further, electing an artificial genetic structure which, say, eliminates age-related diminution of the secretion of human growth hormone and other essential endocrines. On the other hand, a congenitally blind patient might desire, for whatever personal reasons, to retain his blindness. Hence his genetic programmes that result in the blindness phenotype would not, for him, constitute "disease" as long as he fully understands the options and outcomes that are available to him. Whether the broad pool of volitional human phenotypes will tend to converge or diverge is unknown, although the most likely outcome is probably a population distribution with a tall, narrow central peak but with longer tails.

One minor flaw in the volitional normative model of disease is that it relies upon the ability of patients to make fully informed decisions concerning their own physical state. The model crucially involves desires and beliefs, which can be irrational, especially during mental illness, and people normally vary in their ability to acquire and digest information. Patients also may be unconscious or too young, whereupon default standards might be substituted in some cases.

Nevertheless, the volitional normative view of disease appears most appropriate for nanomedicine because it recognizes that the era of molecular control of biology could bring considerable molecular diversity among the human population. Conditions representing a diseased state must of necessity become more idiosyncratic, and may progressively vary as personal preferences evolve over time. Some patients will be more venturesome than others — "to each his own." As an imperfect analogy, consider a group of individuals who each take their automobile to a mechanic. One driver insists on having the carburetion and timing adjusted for maximum performance another driver prefers optimum gas mileage; still another prefers minimizing tailpipe emissions; and yet another requires only that the engine be painted blue. In like manner, different people will choose different personal specifications. One can only hope that the physician will never become a mere mechanic even in an era of near-perfect human structural and functional information; an automobile conveys a body, but the human body conveys the soul. Agrees theorist Guttentag: "The physician-patient relationship is ontologically different from that of a maintenance engineer to a machine or a veterinarian to an animal."

TREATMENT METHODOLOGY

The availability of advanced nanomedical instrumentalities should not significantly alter the classical medical treatment methodology, although the patient experiences and outcomes will be greatly improved. Treatment in the nanomedical era will become faster and more accurate, efficient, and effective. In clinical practice, patient treatment customarily includes up to six distinguishable phases: examination, diagnosis, prognosis, treatment, validation, and prophylaxis. Let us consider each of these, in turn.

The first step in any treatment process is the examination of the patient, including the individual's medical history, personal functional and structural baseline, and current complaints. In classical medicine, interview and observation have long been the cornerstone of examination. In ancient times this was limited to obvious manifestations and simple constellations of observables, such as the Hippocratic facies, the four signs of inflammation noted by Celsus, or pulse rate and fever. Clinicians recognize that the traditional taking and interpreting of oral medical histories from new patients is a subtle and complex art, although some aspects of this process might be automated using voice recognition and text preinterpretation software, somewhat easing the physician's burden.

Advancing technology has also brought a plethora of tests that contribute to accurate diagnosis, including auscultation, microscopy and clinical bacteriology in the 19th century, and radiological scanning, clinical biochemistry, genetic testing, and minimally invasive exploratory surgery in the 20th century.

In the 21st century, new tools for nanomedical testing and observation will include clinical in vivo cytography; real-time whole-body microbiotic surveys; immediate access to laboratory-quality data on the patient, physiological function and challenge tests; tissue composition including direct organelle counts in specified tissue populations; quantitative flowcharts of in cyto secondary messenger molecules, extracellular hormones and neuropeptides; per-compartment cytoglucose inventories; and so forth. Before a proper diagnosis can be made, the physician must also establish the patient's personal functional and structural baseline against which any deviations can be noted and corrected, in keeping with the volitional normative model of disease.

By way of introduction, it is instructive to think about a trivial class of test procedures that might be used to diagnose a simple infectious disease at several different levels of technological competence. Let us consider a patient who presents with signs and symptoms that are nonspecific in nature but which suggest an infectious process — e.g. nasal congestion, mild fever, discomfort and cough. The initial signs are due in part to the body's inflammatory response and in part to the infectious agent itself. The diagnostic goal is to identify the infectious agent.

In the late 20th century, the usual procedure would be to culture a sample taken from the patient, in the microbiology laboratory, using various broths, petri plates, and biochemical tests. Some infectious agents are easy to demonstrate. Beta hemolytic streptococci from a throat swab will grow overnight on a blood agar plate, and colony counts for *E. coli* in a urine sample are available in 24 hours. A throat culture that the lab reports as a mixed culture causes no excitement, and a single isolate of *Staphylococcus epidermidis* in a blood culture is usually regarded as a skin contaminant.

How might nanomedicine handle this test? In the nanomedical era, taking and analyzing microbial samples will be much simpler for the practitioner. Such analysis will be as quick and convenient as the electronic measurement of body temperature using a tympanic thermometer in a late 20th-century clinical office or hospital. The physician faces the patient and pulls from his pocket a lightweight handheld device resembling a pocket calculator. He unsnaps a self-sterilizing cordless pencil-sized probe from the side of the device and inserts the business end of the probe into the patient's opened mouth in the manner of a tongue depressor. The ramifying probe tip contains billions of nanoscale molecular assay receptors mounted on hundreds of self-guiding retractile stalks. Each receptor is sensitive to one of thousands of specific bacterial membrane or viral capsid ligands. An acoustic echolocation transceiver provides gross spatial mapping. The patient says "Ahh," and a few seconds later a three-dimensional colour-coded map of the throat area appears on the display panel that is held in the doctor's hand. A bright spot marks the exact location where the first samples are being taken. Underneath the colour map scrolls a continuously updated microflora count, listing in the leftmost column the names of the ten most numerous microbial and viral species that have been detected, key biochemical marker codes in the middle column, and measured population counts in the right column. The number counts flip up and down a bit as the physician directs probe stalks to various locations in the pharynx to obtain a representative sampling, with special attention to sores or any signs of exudate. After a few more seconds, the data for two of the bacterial species suddenly highlight in red, indicating the distinctive molecular signatures of specific toxins or pathological variants. One of these two species is a known, and unwelcome, hostile pathogen. The diagnosis is completed, the infectious agent is promptly exterminated, and a resurvey with the probe several minutes afterwards reveals no evidence of the pathogen.

DIAGNOSIS

Diagnosis is the determination of the cause and nature of a disease in order to provide a logical basis for treatment and prognosis. Traditionally the diagnostic process begins with a thorough history taken from the patient and a relevant physical examination. Often this sufficed to make a confident diagnosis, but the cause of some illnesses remained uncertain without recourse to additional information such as blood tests or radiological examinations. Nanotechnology-based diagnosis will consist principally of examining the patient to determine how he or she deviates from autogenous reference structures and functions, and then interpreting those deviations as healthy or unhealthy for that patient.

In the 20th century, diagnoses frequently involved a high degree of uncertainty, largely due to the general lack of comprehensive molecular diagnostic tools. Thus diagnosis would be guided by statistical analyses; one branch of decision analysis, called utility analysis, even allowed the patient to participate in the decisionmaking process. When the correct decision is unclear, urges one textbook, it is well to remember time-honoured Hippocratic aphorisms such as "first, do no harm" and "common things occur commonly." The eminent Canadian physician Sir William Osler lamented that "errors of judgement must occur in the practice of an art which consists largely in balancing probabilities." Most doctors would prefer to understand the root cause of medical problems rather than adopt mere statistical approaches.

Nanomedical tools will vastly reduce diagnostic uncertainty. Using nanomedical instrumentalities, doctors will gain access to unprecedented amounts of information about their patients including in-office comprehensive genotyping and real-time whole-body scans for particular bacterial coat markers, tumor cell antigens, mineral deposits, suspected toxins, hormone imbalances of genetic or lifestyle origin, and other specified molecules, producing three-dimensional maps of desired targets with submillimeter spatial resolution. Embedded in vivo nanomedical data archives can provide onboard storage of regularly updated self-diagnostic scans, reducing to a minimum the need for symptomatic interview data from patients who may be unconscious, inarticulate, or verbose, who may have limited powers of self-analysis or self-observation and who may have forgotten, suppressed, or amplified descriptions of symptoms.

Of course, physicians do not require an exhaustive survey of the entire body of each patient to molecular detail to make a valid diagnosis. In any particular case, it is the function of the trained medical mind to quickly ascertain where and where not to look in molecular detail. But in the nanomedical era, powerful tools will be available to allow the practitioner to examine almost any portion of a patient in as much detail as desired, right down to the molecular level, with results available in seconds or minutes, and at reasonable cost.

PROGNOSIS AND TREATMENT

Prognosis is a judgement or forecast, based upon a correct diagnosis, of the future course of a disease or injury, and of the patient's prospects for partial or full recovery. Guttentag identifies prognosis as "the predicted course of the reduced state of the patient's psychosomatic freedom of action," and treatment as "the physician's ability to intervene."

But prognosis is a function of treatment as well as disease. From the post-Hippocratic era through the 18th century, treatments were almost purely empirical and often did more harm than good. During the 19th and early 20th centuries, treatments were scientific but largely homeostatic – the medical intervention was rational but served mainly to assist the body in healing itself. Throughout the remainder of the 20th century, truly curative treatments began to rescue some patients from conditions from which their unaided bodies would not have been able to recover. Although conventional biotechnology will enable some important tissue and cellular replacement treatments by the early 21st century, nanomedicine will enable major reconstructive and restorative procedures at the tissue, cellular and molecular levels and will employ active antibiotic devices. The prognosis will almost always be good, except in cases of severe neural damage and a few other specialized circumstances. Therapeutic treatments will be selected to reverse all pathological effects of disease or injury, with a minimum of pain, discomfort, side-effects, intrusiveness and time, and with a maximum of effectiveness, efficiency, and likelihood of success, though of course some tradeoffs will always exist. Nanomedicine also will excel in the correction of molecular defects of a kind which Nature has no predesigned tools – such as the breakdown and removal of intracellular lipofuscin and the removal of indigestible waste products which interfere with neuronal axon transport.

We may compare the therapeutic response to a simple infection at several different levels of technological competence. Consider a patient who has been diagnosed with eastern equine encephalitis, a mosquito- or tick-borne arbovirus. In the 20th century, there was no specific treatment for this disease. Care was generally supportive, with the doctor attempting to maintain the patient's heart and lung function while the infection ran its course. The prognosis was poor. There was a 50 Percent-75 Percent mortality rate with frequent sequelae including seizures and paralysis, especially in children.

The nanomedical therapy. As before, a nanomedical cure for eastern equine encephalitis may be far simpler, less painful and a great deal quicker. A single therapeutic dose consisting of ~0.1 cm^3 of isotonic saline fluid containing ~10 billion active micron-size virucidal nanodevices, a 10 Percent volumetric nanodevice suspension, is injected into the cerebrospinal fluid. Each therapeutic nanorobot has chemical sensors that can unambiguously recognize fluidborne or in cyto arbovirus particles and, once recognition has occurred, destroy them and also reverse the cellular damage. A nanorobot population of this size should be able to destroy all viral particles and effect needed repairs in at most an hour after which the devices are programmed and equipped either to eliminate themselves from the body or to be manually exfused.

VALIDATION AND PROPHYLAXIS

A proper therapeutic protocol will include a procedure for follow-up to ensure that the prescribed treatment was correctly executed with good results. This step is often neglected in order to save costs and may be considered unimportant by some practitioners because approximately 80 Percent-90 Percent of all illnesses which take patients to the doctor are self-curing or self-limiting. The common cold, most infectious diseases and many minor injuries are problems that usually will resolve on their own even with no treatment. In these cases the purpose of treatment is not to provide a cure, but rather to speed the healing process, improve comfort, and avoid complications. Many nanomedical treatments will require supervision and will run quickly to completion, thus follow-up may come back into vogue. Validation may also be viewed as a post-treatment re-diagnosis to ensure that no disease remains present in the patient.

Prophylaxis is the prevention of disease, typically including patient education, immunization programmes, amelioration of occupational hazards, and other preventive and public health measures. In a treatment environment that is rich in effective antibacterial instrumentalities, those microbes which survive will evolve to produce only modest or negligible symptoms that are insufficiently annoying to motivate a patient to seek professional therapeutic relief. It is well-known that bacteria can modify their behaviour over time. Syphilis had a much more fulminating course in the Middle Ages than it has in the 20th century. Some future strain of the syphilitic microbe might produce negligible symptoms, but we should still insist on its eradication because of its potential to revert to its earlier virulence if allowed to spread unchecked in a more benign form. Preventative procedures may also be needed to discover, diagnose, and treat apparently symptomless diseases, and a variety of molecular-based physiological malfunctions and structural micropathologies may require nanoscale tools in order to detect them.

With medical conditions that require ongoing supervision and adjustment, such as maintaining optimum hormone balance and minimal accumulation of molecular debris, nanoscale monitoring stations may act as onboard cellular guidance systems, stimulating or suppressing endocrine secretion as necessary to preserve an ideal state of equilibrium. In some cases, direct manufacture of compounds not easily produced by ribosomes or other biological organelles may be required.

THE PHYSICIAN-PATIENT RELATIONSHIP

Many other aspects of the physician-patient relationship, especially as this relationship may evolve in the coming era of nanomedicine, are important and worthy of extensive discussion. One such issue is the obligation of both parties in the partnership to tell the truth. The patient as a fellow human being has every right to know the truth about his or her biological condition, but other considerations may enter into the fulfillment of this obligation. Patients have a quite natural anxiety about their own possible death, and it has been claimed that this anxiety implies that no one can be truly objective toward his or her own body. Guttentag observes that "telling an unwelcome truth to the unprepared is as ill-conceived as trying to hide the truth from the prepared."

In the nanomedical era, the sheer number of "truths" that may become available for disclosure will increase enormously even as the terminal prognosis becomes rare. Each human being is believed to possess at least 4-10 potentially serious genetic defects; up to 1 Percent of human DNA is of exogenous viral origin, and as much as 10 Percent of the genome consists of transposons, discrete sequences that are positionally mobile among the chromosomes. Should something be done about this, or not? What should the average patient make of the news that his physician has discovered exactly 57 submicron-scale lamellar defects scattered throughout the compact bone of the caudal epiphysis of the patient's right humerus? In the nanomedical era, people will gain the ability to specify their own physical structure to minute detail, but many patients will not be ready, willing, or able to assume responsibility for this knowledge. Thus there is no ideal substitute for the doctor's interpretative abilities and judgements on the patient's behalf as to the personal significance of specific diagnostic information. As the great clinician Thomas Addis observed in another context: "Honesty with patients requires thought and discipline and effort."

BIOTECHNOLOGY AND MOLECULAR NANOTECHNOLOGY

Some pragmatic readers may be wondering what is so special about molecular nanotechnology, that existing or anticipated biotechnology could not accomplish just as well? After all, biotechnology is already an established medical capability. It has real applications and real products already on the market. Reflecting upon the future possibility of sophisticated mechanical medical nanorobots equipped with powerful nanocomputers, in 1989, one well-known cryobiologist mused that "what is not clear is just what need we would have for such devices." The simplest answer, is that each of the three contemporary branches of "nanotechnology" offers something of unique value to the practice of medicine.

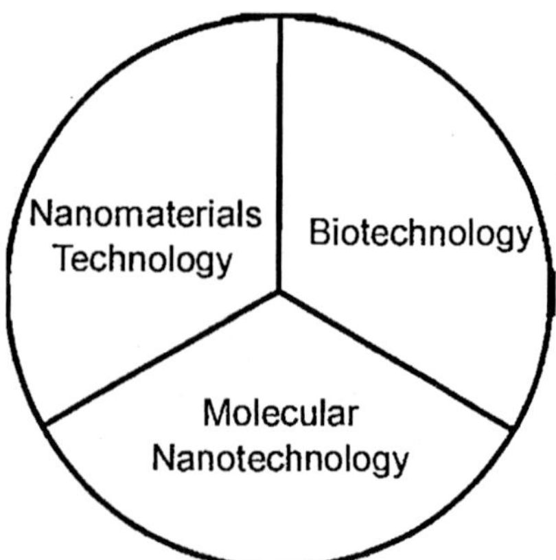

Fig. Three contemporary branches of "nanotechnology"

Nanoscale materials technology has already found widespread use in medicine, including biocompatible materials and analytical techniques, surgical and dental practice, nerve cell research using intracellular electrodes, biostructures research and biomolecular research using near-field optical microscopy, scanning-probe microscopy and optical tweezers, and vaccine design, and also many 20th century bulk chemical and biochemical manufacturing techniques along with much of classical pharmacology.

As for "biotechnology," the original meaning of this word contemplated "the application of biological systems and organisms to technical and industrial processes". In recent times, the field has expanded to include genetic engineering and now takes as its ultimate goal no less than the engineering of all biological systems, even completely artificial organic living systems, using biological instrumentalities.

The third branch, molecular nanotechnology, takes as its purview the engineering of all complex mechanical systems constructed from the molecular level – potentially offering new tools for medical practice. Observes G.M. Fahy, "the difference between nanotechnologists and biotechnologists is that the former do not restrict themselves to the biological limitations of the latter, and they are much more ambitious about the kinds of accomplishments that they want to achieve."

Doctors can utilize solutions to medical problems from all three approaches. As noted earlier, 80-90 Percent of medical complaints resolve themselves via natural homeostatic processes, or without the necessary involvement of active biotechnological or molecular-nanotechnological agents. But by employing biotechnology, the range and efficacy of treatment options greatly increases. With molecular nanotechnology, the range, efficacy, comfort and speed of possible medical treatments again expands enormously. Molecular nanotechnology is essential when the damage to the human body is extremely subtle, highly selective, or time-critical, or when the damage is very massive, overwhelming the body's natural defences and repair mechanisms.

At every difficulty level, most classes of medical problems may be resolved with varying efficacy within the homeostatic/ nanomaterials, biotechnological, or molecular-nanotechnological approaches. As the chosen technology becomes more precise, active, and controllable, the range of options broadens and the quality of the options improves. Thus the question is not whether molecular nanotechnology is required to accomplish a given medical objective. In many cases, it is not though of course there are some things that only biotechnology and molecular nanotechnology can do, and some other things that only molecular nanotechnology can do. Rather, the important question is which approach offers a superior solution to a given medical problem, using any reasonable metric of treatment efficacy. For virtually every class of medical challenge, a mature molecular nanotechnology offers a wider and more effective range of treatment options than any other approach.

It is quite possible to imagine an advanced biotechnology that uses an engineered white cell, fibroblast, or macrophage chassis, energized by native oxygen and glucose and modified mitochondrial powerplants, driven by pseudopodia, cilia or flagella, communicating and navigating via biochemical signals, and even incorporating onboard digital biocomputers to make microscopic biorobots. Principal arguments favouring the biotechnology approach for medical purposes are:

1. That we are already somewhat familiar with such systems, after half a century of intensive molecular biology research;
2. That we have already "built" precursor systems, such as a whole living amoeba constructed from five distinct parts, bioengineered viruses and bacteria as DNA insertion devices, and natural replication stimulated in genetically engineered starter microbes;
3. That biocompatibility will not be a major issue since fibroblasts could be used as the starter material;
4. That both engineered viruses and bacteria are already in wide commercial and research use; and

5. The greater complexity of self-repair in mechanical systems, should it be needed. Many believe that the development pathway to early biorobots may be considerably shorter than for the mechanical nanorobots of the molecular nanotechnology approach, for which in 1998 not a single working prototype yet existed, even in research laboratories.

It is also possible to imagine a molecular nanotechnology that uses mechanical nanorobotic systems. Such systems will have many constitutional differences from biological-based systems. Mechanical systems will transport parts, materials, energy and instructions via fixed channels, whereas most biological systems operate by diffusion. Mechanical systems will have structures constrained by specific geometries, whereas biological systems have structures defined by patterns of containment and interconnection – the shape of a membrane compartment in a cell matters less than its continuity and the contents of the volume it defines. Mechanical systems will be deterministically manufactured by operations analogous to manual construction, whereas engineered ribosomes will self-assemble via diffusion and stochastic matching of complementary parts; in other words, biology uses recipes, while mechanical systems use blueprints. Cells grow, with their parts adapting to one another; mechanical nanorobots may be constructed from parts of fixed structure. Biology uses self-repair; mechanical systems generally do not – largely because self-repair will not be needed in molecular mechanical systems, whose designs may be made more simple by relying upon high component redundancy. These and other differences imply a number of important advantages that mechanical-based medical systems will enjoy over biological-based medical systems, which, taken together, strongly suggest that predominantly mechanical nanosystems may be the approach of choice for a mature medical nanotechnology. The advantages of molecular nanotechnology are many:

1. *Speed of Medical Treatment.* Doctors may be surprised by the incredible quickness of nanorobotic action

when compared to the speeds available from fibroblasts or leukocytes. Normal homeostatic processes such as dermal wound repair via natural fibroblasts may require weeks to run to completion. Typical fibroblast movements occur at 0.1-1 microns/sec but mechanical nanomanipulators can operate at 1-10 cm/sec speeds or faster, a speed advantage of 4-5 orders of magnitude. Even the strongest biological fibres have a failure strength 3 orders of magnitude below the strongest mechanical fibres. Biological cilia beat at ~30 Hz while mechanical nanocilia may cycle up to ~20 MHz, though practical power restrictions and other considerations may limit them to the ~10 KHz range for most of the time. Flexible mechanical surfaces can complete a morphing motion in ~0.1 millisec, compared to the ~100 millisec snapback time for pinched red cell membrane, again a thousandfold advantage in speed. Thus we expect that mechanical therapeutic systems can reach their targets up to ~10,000 times faster, all else equal, and treatments which require ~10^5 sec for a biological system may need only ~10^2 sec for a mechanical system; tachyiatria improves both patient and physician comfort. In either biological or mechanical systems, large numbers of devices of comparable physical size may be employed to do the work, so numbers alone cannot offset the mechanical speed advantage.

2. *Power Density and Transduction.* Biological cells typically employ power densities of 10^3-10^4 W/m^3, with maximum densities of ~10^6 W/m^3 in honeybee flight muscle cells and bacterial flagellar motors By contrast, nanomechanical power systems can produce power densities of 10^9-10^{12} W/m^3, an advantage of 10^3-10^8 for mechanical over biological systems. By 1998, conducting polymer-based actuators generated 20-100 times the force for a given cross-sectional area as mammalian skeletal muscle. Additionally, amoebic locomotion in motile cells requires diffusion-limited

cytoskeletal disassembly and reassembly to achieve movement; mechanical motility systems may employ simple cable-pulling, winches, or ratchets, which are faster and more direct. Some biological energy transducers are reversible, but muscle contraction is irreversible – not only cannot muscle actively re-expand, but stretching it doesn't make it produce much useful chemical energy. In contrast, electric motors can be run backwards to generate electricity, forcing pistons makes them pump, and loudspeakers can be used as microphones.

3. *Superior Building Materials.* Typical biological materials have tensile failure strengths in the 10^6-10^7 N/m^2 range, with the strongest biological materials such as wet compact bone having a failure strength of ~10^8 N/m^2, all of which compare poorly to ~10^9 N/m^2 for good steel, ~10^{10} N/m^2 for sapphire, and ~10^{11} N/m^2 for diamond and carbon fullerenes, again showing a 10^3-10^5 advantage for mechanical systems that use nonbiological materials. Nonbiological materials can be much stiffer, permitting the application of higher forces with greater precision of movement, and they also tend to remain stable over a larger range of temperature, pressure, salinity and pH. Proteins are heat sensitive in part because much of the functionality of their structure is due to noncovalent bonds involved in folding, which are broken more easily at higher temperatures; in diamond, sapphire, and many other rigid materials, structural shape is covalently fixed, hence is far more temperature-stable. Most proteins tend to become dysfunctional at cryogenic temperatures, unlike diamond-based mechanical structures. Biomaterials are not ruled out for all nanomechanical systems, but represent only a small subset of the materials that can be used in nanorobots. Mechanical systems can employ a wider variety of atoms and molecular structures in their design and construction, with novel functional forms

that might be difficult to implement in a biological system such as steam engines or nuclear power. As another example, an application requiring the most effective bulk thermal conduction possible should use diamond, the best conductor available, not some biomaterial with inferior thermal performance.

4. *Nondegradation of Treatment Agents.* Diagnostic and therapeutic agents constructed of biomaterials generally are biodegradable in vivo, although there is a major branch of pharmacology devoted to designing drugs that are moderately non-biodegradable — anti-sense DNA analogues with unusual backbone linkages and peptide nucleic acids are difficult to break down. However, suitably designed nanorobotic agents constructed of nonbiological materials are not biodegradable. An engineered fibroblast may not stimulate an immune response when transplanted into a foreign host, but its biomolecules are subject to chemical attack in vivo by free radicals, acids, and enzymes. Even "mirror" biomolecules or "Doppelganger proteins" comprised exclusively of unnatural D-amino acids have a lifetime of only ~5 days inside the human body. Nonbiological materials such as diamond and sapphire are highly resistant to chemical breakdown or leukocytic degradation in vivo, and pathogenic biological entities cannot easily evolve useful attack strategies against these materials.

5. *Control of Nanomedical Treatment.* Present-day biotechnological entities are not programmable and cannot be switched on and off conditionally during task execution. A digital biocomputer, while possible in theory, represents a considerable conceptual departure from the usual biological paradigm. Even assuming that a digital biocomputer could be installed in a fibroblast, and that appropriate effector mechanisms could be attached, such a system would necessarily have slower clock cycles, less capacious

memory per unit volume, and longer data access times, implying less diversity of action, poorer control, and less complex executable programmes than would be available in nanoscale electromechanical computer systems. The mechanical approach emphasizes precise control of action, including control of physical placement, timing, strength, structure, and interactions with other entities. The biological approach emphasizes the use of poorly controlled natural structures, needlessly sacrificing huge blocks of the available functionality and design space.

6. *Nanodevice Versatility.* Mechanical systems can readily incorporate biological elements if necessary, but artificial biological systems can incorporate nonbiological materials such as carbon nanotubes or diamond/sapphire structural elements only with difficulty, in part because biology has a more limited repertoire of "effector" mechanisms. Artificial biological systems cannot easily incorporate nonbiological materials where desired because natural biological assembly methods make no provisions for these materials either in the coded instructions in DNA or in the attachment chemistries. Rebuilding or reconstructing the human body with nonbiological components, or augmentation of human body function with unnatural abilities or features, will be very difficult or impossible to achieve using purely biotechnological means.
7. *Avoiding Overspecialization.* M. Krummenacker notes that one of the most glaring shortcomings of bacteria and other naturally occurring molecular machinery – when viewed as systems subject to further engineering – is the rather limited range of molecular substrates they can utilize. Bacterial enzymes are highly specialized devices, with very narrow substrate specificities. Thousands of different enzymes are needed in each organism, and the substance classes

capable of digestion are limited to some sugars, various amino acids including proteins that can be degraded by excreted proteases, lipids, and a few other smallish oxygen-functionalized carbon molecules such as glycerol and ethanol. Some bacteria can metabolize CO_2 and a handful of aromatic compounds, but there is a vast range of organic chemicals that most bacteria cannot degrade or manufacture. A much smaller set of substantially more general molecular tools can probably be designed using the mechanical approach; mechanosynthesis can fabricate and assemble, or disassemble, a far wider range of molecular structures than are available to the cellular machinery of life.

8. *Faster and More Precise Diagnosis.* The analytic function of medical diagnosis requires rapid communication between the injected devices and the attending physician. If limited to chemical messaging, biotechnology devices will require minutes or hours to complete each diagnostic loop. Nanomachines, with their more diverse set of input-output mechanisms, can outmessage the results of in vivo reconnaissance or testing literally in seconds. Such nanomachines can also run more tests of greater variety in less time. Mechanical nanoinstruments, including molecule-by-molecule disassemblers, will make comprehensive cell mapping and cell interaction analysis possible. Bacterial resistance can be assayed at the molecular level, allowing new treatment agents to more easily be composed, manufactured and immediately deployed.
9. *More Sensitive Response Threshold for High-Speed Action.* Unlike natural systems, an entire population of nanobiotic devices can be triggered globally by just a single local detection of the target antigen or pathogen. The natural immune system takes $>10^5$ sec to become fully engaged after exposure to a systemic

pathogen or other antigen-presenting intruder. A biotechnologically enhanced immune system that can employ the fastest natural unit replication time will require $\sim 10^4$ sec for full deployment post-exposure. By contrast, a nanobiotic immune system can probably be fully engaged in at most two blood circulation times, or $\sim 10^2$ sec.

10. *More Reliable Operation.* Engineered macrophages would probably individually operate less reliably than mechanical nanorobots. Many pathogens, such as *Listeria monocytogenes* and *Trypanosoma cruzi*, are known to be able to escape from phagocytic vacuoles into the cytoplasm; while biotech drugs or cell manufactured proteins could be developed to prevent this, nanorobotic trapping mechanisms can be more secure. Proteins assembled by natural ribosomes typically incorporate one error per $\sim 10^4$ amino acids placed; current gene and protein synthesizing machines utilizing biotechnological processes have similar error rates. A molecular nanotechnology approach will improve error rates by at least a millionfold. Mechanical systems can also incorporate sensors to determine if and when a particular task needs to be done, or when a task has been completed. Finally, it is unlikely that natural organisms will be able to infiltrate mechanical nanorobots or to co-opt their functions. By contrast, a biological-based robot could be diverted or defeated by microbes that can piggyback on its metabolism, interfere with its normal workings, or even incorporate the device wholesale into their own structures, causing the engineered biomachine to perform some new or different function than was originally intended. There are many examples of such co-option among natural biological systems, including the protozoan mixotrichs found in the termite gut that have assimilated bacteria into their bodies for use as motive engines, and the nudibranch mollusks that steal nematocysts away

from coelenterates such as jellyfish and incorporate the stingers as defensive armaments in their own skins, a process which S. Vogel has called "stealing loaded guns from the army."

11. *Verification of Progress and Treatment.* Using a variety of communication modalities, nanorobots can report back to the attending physician, with digital precision, a summary of diagnostically- or therapeutically-relevant data describing exactly what was found, and what was done, and what problems were encountered, in every cell visited. A biological-based approach relying upon chemical messaging is necessarily slow with limited signaling capacity. Also unlike mechanical nanorobots, biotechnological systems generally cannot monitor their own functions while working, and, except for a few highly specialized DNA proofreading systems, cannot directly inspect their work while it is in progress or after it is finished.
12. *Minimum Side Effects.* Almost all drugs have side effects, such as conventional cancer chemotherapy which causes hair loss and vomiting, although computer-designed drugs have high specificity and relatively few side effects. Carefully tailored cancer vaccines under development in the late 1990s were expected unavoidably to affect some healthy cells. Even well-targeted drugs are distributed to unintended tissues and organs in low concentrations, although some bacteria can target certain organs fairly reliably without being able to distinguish individual cells. By contrast, mechanical nanorobots may be targeted with virtually 100 Percent accuracy to specific organs, tissues, or even individual cellular addresses within the human body. Such nanorobots should have few if any side effects, and will remain safe even in large dosages because their actions can be digitally self-regulated using rigorous control

protocols that affirmatively prohibit device activation unless all necessary preconditions have been, and continuously remain, satisfied. G.M. Fahy has noted that these possibilities could transform "drugs" into "programmable machines with a range of sensory, decision-making, and effector capabilities [that] might avoid side effects and allergic reactions...attaining almost complete specificity of action....Designed smart pharmaceuticals might activate themselves only when, where, and if needed." Additionally, nanorobots may be programmed to excuse themselves from the site of action, or even from the body, after a treatment is completed; by contrast, spent biorobotic elements containing ingested foreign materials may have more limited post-treatment mobility, thus lingering at the worksite causing inflammation when naturally degraded or removed.

13. *Reduced Replicator Danger*. Drexler points out that living systems are evolved systems, while nanomechanical replicators would be designed: "The former are shaped to serve the goal of their own survival and replication in a natural environment, whereas the latter will be shaped to serve human goals, perhaps in an artificial environment." Genetic engineering involves not design of replicators from scratch, but tinkering with the molecular machinery of existing bioreplicators. Since bioreplicators were not designed, they are not necessarily structured in ways that lend themselves to complete understanding, and processes based on diffusion and matching allow complex nonlocal interactions that can be hard to trace. Bioreplicators can be crippled, but having evolved in nature, they resemble systems that can survive in nature. Typically, they are able to exchange genetic information with wild organisms, raising the possibility of the introduction of new, unconstrained replicators in the natural environment.

Having evolved to evolve, they have a capacity for further evolution – to serve their own survival, not human goals. Even when stripped of key pieces of DNA to interfere with its replication powers, a live attenuated AIDS vaccine can slowly recover its virulence and can attack immune cells. R. Bradbury suggests that artificial biorobots could incorporate multiple fail-safe mechanisms including required external essential nutrients or suicide suppressors, self-destruct triggers, countdown timers like telomeres, and engineering to reject foreign DNA, but it remains logically easier for a system with inchoate capacity to evolve to resume doing so, than for a system which has never had this capability to spontaneously develop it.

In contrast, nanomechanical replicators will be designed from scratch and thus will differ fundamentally from biological systems. The parts and structures of designed mechanical systems will be known, and the relationships among their parts will also be designed and fixed. More important, nanoreplicators will be fundamentally alien to the biosphere, unrelated to anything that has evolved to survive in nature. Certainly the capacity to fail can appear by accident, and emergent capabilities cannot be completely ruled out, but engineering experience shows that the ability to perform complex organized activities does not normally appear spontaneously. Also, and purely as a geometrical consideration, adding a new part inside a densely organized geometric structure typically requires changes in the relative positions of many other parts, and hence corresponding adjustments in design. Adding a part inside a densely organized topological structure typically leaves topologies unchanged – room can be made by stretching and shifting other parts, with no change in their essential design, hence permitting easier modification to biological structures by exogenous agencies.

Note that mechanical medical nanodevices need not be capable of replication. There is no requirement for replication in vivo; such replication would be needlessly dangerous, and adding this capability would reduce effectiveness in carrying out the primary medical task. Analogously, viral vectors employed in genetic therapies are modified to be "incapable" of replication.

14. *Assured Patentability*. Microscopic biorobots, unavoidably derived from natural biological material, may someday be deemed unpatentable under a general prohibition on "genetic colonialism" or other emerging legal doctrines. In contrast, mechanical nanorobots, being fully-artificial and designed machines, should always be patentable provided they satisfy the customary legal criteria

3

Nanomedicine Taxonomy

Nanomedicine research is being funded by government sources, such as the National Institutes of Health, and by companies in various sectors, including pharmaceutical, biotechnology and medical devices. Because of the focused and shorterterm development horizon of commercial applications, nanomedicine research is more easily characterized in these areas.

The nanomedicine taxonomy summarized below and further described in this document, classifies some of the leading areas that nanotechnology tools, materials, devices, and intelligent materials and machines are currently applied in medical research. Specific company and university research activities should be interpreted as examples and are only a subset of the vast scope of nanomedicine research currently underway.

Nanomedicine Taxonomy

Biopharmaceutics

- Drug Delivery
  - Drug Encapsulation
  - Functional Drug Carriers
- Drug Discovery

Implantable Materials

- Tissue Repair and Replacement
  - Implant Coatings
  - Tissue Regeneration Scaffolds

Structural Implant Materials

- Bone Repair
- Bioresorbable Materials
- Smart Materials

Implantable Devices

- Assessment and Treatment Devices
  - Implantable Sensors
  - Implantable Medical Devices
- Sensory aids
  - Retina Implants
  - Cochlear Implants

Surgical Aids

- Operating Tools
  - Smart Instruments
  - Surgical Robots

Diagnostic Tools

- Genetic Testing
  - Ultra-sensitive Labeling and Detection and Technologies
  - High Throughput Arrays and Multiple Analyses

BIOPHARMACEUTICS

Drug Delivery

Nanotechnology provides a wide range of new technologies for developing customised solutions that optimize the delivery of pharmaceutical products. To be therapeutically effective, drugs need to be protected during their transit to the target action site in the body while maintaining their biological and chemicals properties. Some drugs are highly toxic and can cause harsh side effects and reduced therapeutic effect if they decompose during their delivery. Depending on where the drugs will be absorbed, and whether certain natural defence mechanisms need to be passed through such as the blood-brain barrier, the transit time and delivery challenges can be greatly different.

Once a drug arrives at its destination, it needs to be released at an appropriate rate for it to be effective. If the drug is released too rapidly it might not be completely absorbed, or it might cause gastro-intestinal irritation and other side effects. The drug delivery system must positively impact the rate of absorption, distribution, metabolism, and excretion of the drug or other substances in the body. In addition, the drug delivery system must allow the drug to bind to its target receptor and influence that receptor's signalling and action, as well as other drugs, which might also be active in the body.

Drug delivery systems also have severe restrictions on the materials and production processes that can be used. The drug delivery material must be compatible and bind easily with the drug, and be bioresorbable. The production process must respect stringent conditions on processing and chemistry that won't degrade the rug, and still provide a cost effective product.

Nanotechnology can offer new drug delivery solutions in the following areas.

Drug Encapsulation

One major class of drug delivery systems is materials that encapsulate drugs to protect them during transit in the body. Drug encapsulation materials include liposomes and polymers which are used as microscale particles. The materials form capsules around the drugs and permit timed drug release to occur as the drug diffuses through the encapsulation material. The drugs can also be released as the encapsulation material degrades or erodes in the body. When encapsulation materials are produced from nanoparticles in the 1 to 100nm size range instead of bigger microparticles, they have a larger surface area for the same volume, smaller pore size, improved solubility, and different structural properties.

This can improve both the diffusion and degradation characteristics of the encapsulation material. In addition to liposomes and polymers, other types of nanoparticles are available for encapsulation. Materials such as silica and calcium phosphate have demonstrated superior properties at the nanoscale than the microscale, and can potentially be better suited for certain drug delivery challenges mentioned above. Nanoparticle encapsulation is also being investigated for the treatment of neurological disorders to deliver therapeutic molecules directly to the central nervous system beyond the blood-brain barrier, and to the eye beyond the blood-retina barrier. Applications could include Parkinson's, Huntington's, Alzheimer's, ALS and diseases of the eye.

In particular, ophthalmic disorders represent a rapidly growing area that is threatened by various diseases such as age related macular degeneration, diabetic retinopathy, glaucoma, and retinitis pigmentosa. Neurotech is developing an encapsulated cell therapy to treat eye diseases. It uses a semipermeable membrane to encapsulate cells, which also permits therapeutic agents produced by the cells to diffuse through the membrane. The membrane isolates the cells from the local environment and minimizes immune rejection. The encapsulated cells are administered by a device implanted in the eye to permit the continuous release of therapeutic molecules from living cells. This avoids direct injections into the eye, which may not be practical for regular administrations. Advectus Life Sciences is developing a nanoparticle-based drug delivery system for the treatment of brain tumors.

The anti-tumor drug doxorubicin is adhered to a nanopolymer Poly ButylCyano Acrylate particle and coated with polysorbate 80. The drug is injected intravenously and circulates through the blood stream. The polysorbate 80 attracts plasma apolipoproteins and is used by the blood stream to carry lipids. This is intended to create a camouflage effect similar to LDL cholesterol, which allows the drug to pass the blood-brain barrier.

Functional Drug Carriers

Another class of drug delivery systems where nanotechnology offers interesting solutions is in the area of nanomaterials that carry drugs to their destination sites and also have functional properties. Certain nanostructures can be controlled to link with a drug, a targeting molecule, and an imaging agent, then attract specific cells and release their payload when required. Because of their nanoscale size, nanostructures have the ability to enter the cells, as cells will typically internalize materials below 100nm. Some of the leading nanostructures being used for this purpose include fullerenes, dendrimers, and nanoshells.Fullerenes are natural hollow spheres, one nanometer in diameter, made up of 60 carbon atoms.

Fullerenes create a unique drug delivery platform that allows active pharmacopheres to be grafted to its surface in three-dimensional orientations for precise control in matching fullerene compounds to biological targets, in entrapping atoms within the fullerene cage, and for attaching fullerene derivatives to targeting agents. C Sixty is developing fullerene-based drug delivery platforms which link fullerenes with antibodies and other targeting agents. Some of C Sixty's drug delivery systems include fullerene-decorated chemotherapeutic constructs, fullerene-radiopharmaceuticals, and fullerene-based liposome systems, called Buckysomes, for the delivery of single drug loads or multiple drug cocktails. Employing rational drug design, C Sixty has produced several drug candidates using its fullerene platform technology in the areas of HIV/AIDS, neuro-degenerative disorders and cancer. Another nanomaterial used as a drug delivery scaffold is the dendrimer, a polymer molecule discovered by Don Tomalia of Dendritic Nanotechnologies.

Researchers such as James Baker of the University of Michigan are using dendrimers to get genetic material or tumor-destroying therapies into a cell without triggering an immune response. This is due to the dendrimer's small size and branched structure. Dendrimers can be designed to release attached compounds in response to a specific molecule or chemical reaction. A layered sphere called a nanoshell is being developed by Nanospectra for drug delivery. The nanoshell has a gold exterior layer which covers interior layers of silica and drugs. Nanoshells can be made to absorb light energy and then convert it to heat. As a result, when nanoshells are placed next to a target area such as tumor cell, it can release tumorspecific antibodies when infrared light is administered.

Drug Discovery

Nano and micro technologies are part of the latest advanced solutions and new paradigms for decreasing the discovery and development times for new drugs, and potentially reducing the development costs. Traditional trial-and-error methods have contributed to a discovery process lasting 10 years or more for new drugs to reach the market.

In recent years, a number of new and complementary technologies have been developed which considerably impact the drug discovery process. High-throughput arrays and ultra-sensitive labeling and detection technologies are being used to increase the speed and accuracy of identifying genes and genetic materials for drug discovery and development. These micro and nano technologies along with information technology solutions such as combinatorial chemistry, computational biology, computer-aided drug design, data mining, and data processing tools are addressing the challenges related to eliminating critical bottlenecks in drug discovery. In addition to the capacity for processing information, the number of drug candidates that have been screened in the last ten years has increased by three orders of magnitude from 500,000 drug compounds to approximately 1.5 billion according to BCC.

IMPLANTABLE MATERIALS

Tissue Repair and Replacement

Nanotechnology provides a new generation of biocompatible nanomaterials for repairing and replacing human tissues. Human tissue that is diseased or traumatically compromised may require synthetic materials for its repair or replacement. While most types of tissues repair the interaction of stem cells with chemical modulators, there are differences in the ways that various tissues heal.

"Hard" tissues such as bone and teeth heal by reproducing tissues indistinguishable from the original. However in cases where a dental or artificial bone implant is required, the structural material used in the implant may trigger immune rejection, corrode in the body fluids, or no longer bond to the host bone. This can require additional surgery or result in the loss of the implant's function. In many cases, the failure occurs at the tissue-implant interface, which may be due to the implant material weakening its bond with the natural material. To overcome this, implants are often coated with a biocompatible material to increase their adherence properties and produce a greater surface area to volume ratio for the highest possible contact area between the implant and natural tissue.

"Soft" tissues such as skin, muscle, nerves, blood vessels and ligaments repair damaged areas with fibrous tissue. Damaged tissue from various sources such as burns and ulcers can be self-repaired by the body, but can also result in scar formation. Graft material using artificial sheets can replace skin and other tissue with reasonable graft stability and cosmetic outcome.

In other types of tissue, notably "Ultrasoft" tissue such as cell membrane and organelles that exhibit metabolic function, tissue replacement can best occur when living cells are transplanted in a mesh-structured synthetic scaffold. The scaffold incorporates signalling ligands or DNA fragments to elicit specific cellular responses, and molecular sensors to accept feedback from the in vivo environment. The scaffold is typically a temporary structure that is bioresorbable when the tissue is regenerated. The scaffold material needs to be fabricated into a desired shape or threedimensional structure, with specific surface properties to support the site where the organized growth of multiple cell types will take place. Nanotechnology can new offer new solutions for tissue repair and replacement in the following areas.

Implant Coatings

Nanotechnology brings a variety of new high surface area biocompatible nanomaterials and coatings to increase the adhesion, durability and lifespan of implants. Ceramic materials such as calcium phosphate are made into implant coatings using nano-sized particles instead of micro-sized particles. In addition to the higher surface areas and improved adhesion properties of the nanoparticle coatings, improved coating techniques are also being developed.

While high temperature processes such as plasma spray can melt ceramic particles and reduce their surface area and adhesion properties, new low temperature processes with electromagnetic fields can maintain the nanomaterial properties. This provides the maximum possible contact area between the implants and bone surface to improve the potential for in-growth in the host bone. New types of nanomaterials are being evaluated as implant coatings to improve interface properties. Nanopolymers such as polyvinyl alcohol can be used to coat implantable devices that are in contact with blood for dispersing clots or preventing their formation.

Tissue Regeneration Scaffolds

Nanostructures are being researched for the preparation and improvement of tissue regeneration scaffolds. Research areas include the ability to develop molecularly sensitive polymers using the optical properties of nanoparticles as control systems, manipulating the stiffness and strength of scaffolds using hybrid nanostructures, and the use of nanotechnology to prepare molecular imprints to maximize long-term viability and function of cells on scaffold surfaces. With the ultimate objective of growing large complex organs, a variety of nanomaterials and nanotechnology fabrication techniques are being investigated as tissue regeneration scaffolds that provide improved structural requirements and guide the activity of seeded cells.

Some examples are as follows:

- Nanoscale polymers such as Polyvinyl alcohol are being molded into heart valves and seeded with fibroblasts and endothelial cells.
- PVA is also being investigated for the cornea by having corneal epithelia cells seeded in a PVA hydrogel structure. This polymer material can absorb more than 20 Percent its weight in water while maintaining a distinct three-dimensional structure.
- A polyglycolic ball is being experimented with muscle cells and bladder endothelial cells.
- Polymer nanocomposites are being researched for bone scaffolds.

Commercially viable solutions are thought to be 5 to 10 years away. Scientific challenges related to a better understanding of molecular/cell biology and fabrication methods for producing large three-dimensional scaffolds are among the many obstacles yet to be overcome. Nanostructures are also being used to study the fundamental properties of implanted tissues. In areas of in vivo analysis, nanostructures are used as tracers for implanted cells, and to study the response of host to implanted tissues.

Structural Implant Materials

Nanotechnology provides a new generation of biocompatible materials that can be used as implants or temporary biosorbable structures. Bone is a high strength material that is used as both weight bearing and non-weight bearing structures. Bones are more than just structural materials as they also contain interconnected pores that allow body fluids to carry nutrients and permit interfacial reactions between hard and soft tissues.

In the case of bone fractures, grafts, disorders, dental applications and other types of surgery, bones may require repair or replacement. A variety of natural materials are used as bone substitutes. These include autograft from the patient's pelvis, allograft from another human, bovine material or coral blocks. Natural materials tend to be brittle and can lose mechanical strength during sterilization. They can also cause inflammation, pain at the pelvis graft site, and potentially transmit disease. Bone cavities can also be filled with synthetic bone cement. Current bone cements containing polymethylmethacryate act as a filler or grout, which is injected as a flowable paste and then hardens in vivo.

While PMMA cement can offer adequate mechanical properties and bonding, it is typically recommended only for non-weight bearing bones. PMMA has also been linked to tissue damage, nerve root pain and other side effects.

Bone Repair

Nanotechnology brings a variety of new high surface area biocompatible nanomaterials that can be used for bone repair and cavity fillers. High strength nanoceramic materials, such as calcium phosphate apatite and hydroxyapatite, can be made into a flowable, moldable nanoparticle paste that can conform to and interdigitate with bone.

As natural bone is approximately 70 Percent by weight CPA including hydroxyapatite, biocompatibility is thought to be extremely high with minimal side effects. As its dense surface and tight three dimensional crystalline structure will allow for a superior compressive strength to PMMA, nanoceramics may be suitable for both weight bearing and non-weight bearing bones. Nanoceramics such as CPA is nanocrystalline and can have grain size of under 50nm. Its nanocrystals can be connected to each other to form connective hard tissue required for bones. Bone cement material being developed at the University of South Carolina and Competitive Technologies, can harden in vivo and in the presence of serum, form a solid bone-like structure capable of stabilizing a fractured bone or used as a bone substitute.

Bioresorbable Materials

Nanotechnology also brings advances in bioresorbable materials. Bioresorbable polymers are currently being used in degradable medical applications such as sutures and orthopaedic fixation devices. With new production methods, nanostructures are being fabricated which could be used as temporary implants. Bioresorbable implants will biodegrade and do not have to be removed in a subsequent operation. In one application, nanostructured implants are being designed to degrade at a rate that will slowly transfer load to a healing bone that it is supporting.

Research is also being done on a flexible nanofibre membrane mesh that can be applied to heart tissue in open-heart surgery. The mesh can be infused with antibiotics, painkillers and medicines in small quantities and directly applied to internal tissues. The nanomaterial will degrade over time and not stick to surgeons' wet gloves, which can complicate the use of some materials currently in use.

Smart Materials

Smart materials are a class of nanomaterials that respond to changes in the environment such as a drop in temperature or pH. An environmental change could trigger a physical or chemical effect that mimics a natural mechanism in the body. Applications could include a smart polymer that flexes with mechanical strength as an artificial muscle, or a hydrogel that dissolves according to body chemistry to more efficiently deliver drugs. When attached to or embedded in a structure, smart materials can potentially sense disturbances, or trigger physical reactions with force in actuators, piezo sensors, and shape memory metals.

It is possible to have smart materials used in the body to exceed current human performance on multiple dimensions, and detect environmental conditions beyond current human limits.

IMPLANTABLE DEVICES

Assessment and Treatment Devices

Nanotechnology offers sensing technologies that provide more accurate and timely medical information for diagnosing disease, and miniature devices that can administer treatment automatically if required. Health assessment can require medical professionals, invasive procedures and extensive laboratory testing to collect data and diagnose disease. This process can take hours, days or weeks for scheduling and obtaining results. Some medical information is extremely time sensitive such as finding out if there is sufficient blood flow to an organ or tissue after transplant or reconstructive surgery, before irreversible damage occurs.

Certain medical tests such as biopsies are subjective and can provide inconclusive or incorrect results. In a false negative result where a needle misses the tumor and then samples a normal tissue, the cancer may go untreated and can impact a patient's chances for long-term survival. Some tests such as diabetes blood sugar levels require patients to administer the test themselves to avoid the risk of their blood glucose falling to dangerous levels. Certain users such as children and the elderly may not be able to perform the test properly, timely or without considerable pain. People who are exposed to radiation or hazardous chemicals in their work environment are at a higher risk of illness. Occasional testing is typically done but may not detect a disease in its early stage.

Early detection could initiate timely treatment with a higher chance of success, and have a worker removed from the hazardous environment to prevent further damage. Nanotechnology can new offer new implantable and/or wearable sensing technologies that provide continuous and extremely accurate medical information. Complementary microprocessors and miniature devices can be incorporated with sensors to diagnose disease, transmit information and administer treatment automatically if required. Example applications are as follows.

Implantable Sensors

Micro and nanosized sensors can make use of a wide range of technologies that most effectively detect a targeted chemical or physical property. Researchers at Texas A&M and Penn State use polyethylene glycol beads coated with fluorescent molecules to monitor diabetes blood sugar levels.

The beads are injected under the skin and stay in the interstitial fluid. When glucose in the interstitial fluid drops to dangerous levels, glucose displaces the fluorescent molecules and creates a glow.

This glow is seen on a tattoo placed on the arm. Another approach by researchers at the University of Michigan is using dendrimers attached with fluorescent tags to sense pre-malignant and cancerous changes inside living cells. The dendrimers are administered transdermally and because of their small size, pass through membranes into white blood cells to detect early signs of biochemical changes from radiation or infection. Radiation changes the flow of calcium ions within the white blood cells and eventually triggers apoptosis, or programmed cell death due to the radiation or infection.

The fluorescent tags attached to the dendrimers will glow in the presence of the death cells when passed with a rentinal scanning device using a laser capable of detecting the fluorescence. A similar application is being researched with NASA for detecting radiation levels in astronauts. Sensor microchips are also being developed to continuously monitor key body parameters including pulse, temperature and blood glucose. A chip would be implanted under the skin and transmit a signal that could be monitored continuously. Another application uses optical microsensors implanted into subdermal or deep tissue to monitor tissue circulation after surgery.

Instrumentation is linked to transmit data to a nearby sensor to provide very early indication of inadequate circulation if the surgery is unsuccessful. Another type of implantable sensor uses MEMS devices and accelerometers for monitoring and treating paralysed limbs. Implantable MEMS sensors can measure strain, acceleration, angular rate and related parameters to determine normal and problem data.

Implantable Medical Devices

Implantable sensors can also work with a series of medical devices that administer treatment automatically if required.

Tiny implantable fluid injection systems can dispense drugs electrically on demand making use of microfluidic systems, miniature pumps, and reservoirs. Initial applications may include chemotherapy that directly targets tumors in the colon and are programmed to dispense precise amounts of medication at convenient times, such as after a patient has fallen asleep. Lupus, diabetes and HIV/AIDS applications are also being investigated. Implantable sensors that monitor the heart's activity level can also work with an implantable defribulator to regulate heartbeats.

These devices are used with microprocessors to deliver electricity that keeps the heart in rhythm when levels go above or below the person's normal heart range. Implantable MEMS devices are also fitted on prostheses to mimic the stability and strain of natural limbs. An artificial leg being developed uses sensors to measure load on the foot, knee angles and motion over 50 times per second. The sensors work with an electronically controlled hydraulic knee to improve its stability. Implantable medical devices that are greater than 1mm in diameter might unfavourably alter the functions of surrounding tissue. Smaller implantable devices with non-intrusive or minimally-intrusive systems will likely contain nanoscale materials and smaller systems approaching the nanoscale. Functional Electrical Stimulation is a method for treating people to regain the use of their paralyzed limbs by electrically stimulating paralyzed muscles with implanted electrodes. Researchers at Aalborg University in Denmark are applying nanostructures to the electrode surfaces to improve biocompatibility and acceptance in the neural/muscle tissue.

When placed within a cell membrane, the nanostructures form a bioelectric interface with the neuron or muscle cell that enables the intracellular potential of the cell to be observed and manipulated. Researchers at Aalborg are also using nanostructures to activate denervated muscles caused by injuries to the lower motor neurons located in the spinal cord. This can result from traumatic spinal cord injury, strokes in the spinal cord, repeated vertebral subluxation, brachial plexus injuries and peripheral myopathies such as polio which destroys the nerve cells controlling muscle. The muscle fibre membrane is incorporated with potential-generating nanostructures that change the transmembrane potential of the muscle fibre and improves the extra cellular electrical stimulation. Muscle fibre activation is achieved by illuminating the incorporation site on the muscle fibre to optically activate denervated muscles.

Sensory Aids

Nano and related micro technologies are being used to develop a new generation of smaller and potentially more powerful devices to restore lost vision and hearing functions. The devices collect and transform data into precise electrical signals that are delivered directly to the human nervous system. Degenerative diseases of the retina, such as retinitis pigmentosa or age related macular degeneration, decrease night vision and can progress to diminishing peripheral vision and blindness.

These retinal diseases may lead to blindness due to a progressive loss of photoreceptors, the light sensitive cells of the eye. In cases where the neural wiring from the eye to the brain is still intact but the eyes' lack photoreceptor activity, photoreceptor loss could be compensated by bridging or bypassing the destroyed photoreceptors and artificially stimulating the adjacent intact cells. Artificially generated impulses could reach the brain and produce visual perception, thereby restoring some elementary vision. In severe hearing loss, patients typically have absent or malfunctioning sensory cells in the cochlea. In a normal ear, sound energy is converted to mechanical energy by the middle ear, which is then converted to mechanical fluid motion in the cochlea.

Within the cochlea, the inner and outer hair sensory cells are sensitive transducers that convert mechanical fluid motion into electrical impulses in the auditory nerve. Cochlear implants are designed to substitute for the function of the middle ear, cochlear mechanical motion, and sensory cells. The implants transform sound energy into electrical energy that will initiate impulses in the auditory nerve. Cochlear implants include an electronic circuit that is surgically placed in the skull behind the ear on the mastoid process of the temporal bone. This circuit is attached to a bundle of tiny wires that are inserted into the cochlea. At the end of the wires are typically 8 to 24 electrodes that cause a different pitch percept when stimulated. The other part of the device is external and has a microphone, a speech processor, and connecting cables.Current cochlear implants have a number of drawbacks.

They require major surgery and can eliminate any remaining natural hearing. Because of their large size, current cochlear implants often stimulate several nerve fibres at once. This causes users to experience imprecise or distorted sound, especially with complex sounds like music. Microtechnologies and eventually nanotechnologies are being used in a new generation of implantable devices for improving sensory loss in the following areas..

Retina Implants

Retinal implants are in development to restore vision by electrically stimulating functional neurons in the retina. One approach being developed by various groups including a project at Argonne National Laboratory is an artificial retina implanted in the back of the retina. The artificial retina uses a miniature video camera attached to a blind person's eyeglasses to capture visual signals.

The signals are processed by a microcomputer worn on the belt and transmitted to an array of electrodes placed in the eye. The array stimulates optical nerves, which then carry a signal to the brain. Another approach by Optobionics makes use of a subretinal implant designed to replace photoreceptors in the retina. The visual system is activated when the membrane potential of overlying neurons is altered by current generated by the implant in response to light stimulation. The implant makes use of a microelectrode array powered by as many as 3,500 microscopic solar cells.

Cochlear Implants

A new generation of smaller and more powerful cochlear implants are intended to be more precise and offer greater sound quality. An implanted transducer is pressure-fitted onto the incus bone in the inner ear. The transducer causes the bones to vibrate and move the fluid in the inner ear, which stimulates the auditory nerve. An array at the tip of the device makes use of up to 128 electrodes, which is five times higher than current devices. The higher number of electrodes provides more precision about where and how nerve fibres are stimulated.

This can simulate a fuller range of sounds. The implant is connected to a small microprocessor and a microphone, which are built into a wearable device that clips behind the ear. This captures and translates sounds into electric pulses which are send down a connecting wire through a tiny hole made in the middle ear. Implant electrodes are continuously decreasing in size and in time could enter the nanoscale.

SURGICAL AIDS

Operating Tools

Medical devices that contain nano and micro technologies will allow surgeons to perform familiar tasks with greater precision and safety, monitor physiological and biomechanical parameters more accurately, and perform new tasks that are not currently done. Surgery can be physically demanding on both the surgeon and the patient. Open surgery such as heart operations can require a surgeon to make a large cut into the skin and underlying muscles, saw through breastbones, crack open the ribcage, then make delicate and high precision maneuvers with various surgical instruments. Not only can the patient incur severe pain, scarring, complications and extensive recovery time, the surgeon must be intensely focused for long hours hunched over an operating table.

This can increase the chance of a surgical error due to fatigue in the neck and back, along with potentially significant hand tremors. In various procedures, the surgeon's view could be limited. Judgment may be required to tell whether the tissue about to be cut has the density of the intended cartilage, or alternatively, a tendon which was grabbed by mistake. To improve surgical results benefiting both the patient and surgeon, minimally invasive surgical procedures are increasingly being done using laparoscopic techniques. Making use of small entry ports into the area of interest, a rod shaped telescope attached to a camera and other long and narrow surgical instruments are used to perform all of the major maneuvers. Even with its considerable benefits, laparoscopic surgery causes a major shift in surgical skills that have to be mastered as the surgeon's vision and access is severely restricted.

Visual cues are obtained from a two dimensional video display instead of a three dimensional operating field. As well, the surgeon must work with long laparoscopic instruments to handle delicate tissues. Nano and micro technologies are being applied to improve the role of the surgeon in the following areas.

Smart Instruments

Surgical tools such as scalpels, forceps, grippers, retractors and drills are being embedded with miniature sensors to provide real-time information and added functionality to aid surgeons. Surgeons can be given continuous data on the force and performance of their instruments, the tissue type about to be cut and specific tissue properties, such as density, temperature, pressure, and electrical impulses. Verimetra is developing an enhanced version of its Data Knife with logic and surgical micro-electromechanical systems that can provide added information and functionality to assist a surgeon during a procedure by stimulating electrodes, measuring and cutting with ultrasonic elements, and cauterizing.

Instruments are being developed with specific functionality like tilt and pressure to allow neurosurgeons to perform operating tasks with greater precision and safety. Nanoparticles are also being investigated for optically guiding surgery. This can potentially allow for better removal of lesioned or diseased sites, including tumors.

Surgical Robotics

Robotic surgical systems are being developed to provide surgeons with unprecedented control over precision instruments. This is particularly useful for minimally invasive surgery. Instead of manipulating surgical instruments, surgeons use their thumbs and fingers to move joystick handles on a control console to maneuver two robot arms containing miniature instruments that are inserted into ports in the patient. The surgeon's movements transform large motions on the remote controls into micro-movements on the robot arms to greatly improve mechanical precision and safety. A third robot arm holds a miniature camera, which is inserted through a small opening into the patient. The camera projects highly magnified 3-D images on a console to give a broad view of the interior surgical site.

The surgeon controlling the robot is seated at an ergonomically designed console with less physical stress than traditional operating room conditions. UCI Medical Center's da Vinci Surgical System is currently approved for gall bladder, prostrate, colorectal, gynecological, esophageal and gastric bypass procedures. Clinical trials are underway for vascular repair and coronary bypass surgery. The Zeus Robotic Surgical System is presently used for cardiac surgery.

DIAGNOSTIC TOOLS

Genetic Testing

Nano and micro technologies provide new solutions for increasing the speed and accuracy of identifying genes and genetic materials for drug discovery and development, and for treatment-linked disease diagnostics products. Thousands of genes and their products function in a complicated and orchestrated manner. Genetic testing technologies used for drug discovery and in diagnostic products make it possible to measure gene sequences and gene expression levels, which are important indicators of growth, metabolism, development, behaviour and adaptation of living systems.

With these tools, many different genes and expression patterns can be characterized as normal or diseased tissue, along with the various stage of disease. There are many steps and processes involved in genetic testing. While the nature of the tests and nomenclature may vary, some of the key principals and related challenges are described below.

Create or Obtain DNA Arrays of "Known" Probe

Known probes of genetic sequences are prepared or obtained commercially for a desired test. These probes are typically in the form of cDNA or an array of oligonucleotides. While it is desirable to have as many probes as possible to maximize the search for possible results, there is currently a limited number of known specimens. This limits the range of experimentation for drug discovery and disease diagnostics.

With advances in the human genome, gene cloning and molecular biology, a vast amount of new probes are increasingly available which could in time cover the entire genome, and potentially screen multiple targets in a single test. DNA arrays are created by placing probes as individual spots in an orderly array arrangement on a glass microscope slide, a nylon membrane substrate or a silicon chip. The latter case is also referred to as a biochip.

Step 2 – Collect and assay "unknown" targets

Unknown cells or molecules of interest need to be collected and then compared with the known probes in the DNA array. This begins with tissue, blood or other biological samples where nucleic acids are isolated. The nucleic acids are converted into labeled targets by tagging the DNA marker with an observable fluorescent or radioactive label. The labeled targets are incubated with the probes to permit hybridization and form stable DNA and/or RNA duplex. This is used for locating or identifying nucleotide sequences of interest. After the incubation, nonhybridized samples are washed away, and measurements are made of the signal produced when hybridization occurs at particular probe locations.

The level of hybridization between a specific probe and a target indicates the level of the gene corresponding to that probe in the test solution. Current assay technologies use fluorescent dyes to label molecules and require expensive equipment such as a laser to light up biological interactions, and an optical microscope to detect the binding sites. Fluorescent dyes are not always precise or sufficiently sensitive to detect every gene. Also, genes that are not distinguished separately can bleed together and result in a false or inconsistent result. The hybridization may also require gene amplification or Polymerase Chain Reaction which is used to produce millions of copies of a specific piece of DNA in a test tube from a single cell. Amplification could require a repetitive series of cycles making use of various chemical and processing stages. Nano and micro technologies can provide new solutions in the following areas.

Ultra-sensitive Labeling and Detection Technologies

Several new technologies are being developed to improve the ability to label and detect unknown target genes. At Genicon, gold nanoparticle probes are being treated with chemicals that cling to target genetic materials and illuminate when the sample is exposed to light. Another approach by Chad Mirkin and Nanosphere uses gold nanoparticle probes coated with a string of nucleotides that complement one end of a target sequence in the sample. Another set of nucleotides, complementing the other end, is attached to a surface between two electrodes.

If the target sequence is present, it anchors the nanoprobes to the surface like little balloons, and when treated with a silver solution, they create a bridge between the electrodes and produce a current. Quantum Dot Corp. uses quantum dots to detect biological material. Because their colour can be tailored by changing the size of the dot, the potential for multiple colours increases the number of biological molecules that can be tracked simultaneously. In addition, quantum dots do not fade when exposed to ultraviolet light and the stability of their fluorescence allows longer periods of observation. These technologies are expected to be more sensitive than fluorescent dyes and could more effectively detect low abundance and low-level expressioning genes.

They may also make use of smaller and less expensive equipment to light and detect the samples. Without the need for gene amplification, they can also provide results in less time.

High Throughput Arrays and Multiple Analyses

Arrays are usually classified by their sample spot sizes with 200-micrometer diameter as the traditional cutoff between macroarrays and microarrays. Microarrays contain thousands of spots. As the sample spot size further decreases in size, more spots can fit onto a substrate. This reduces the area and corresponding time that needs to be sampled. With continued miniaturization beyond micro, the possibility exists to greatly increase the number of spots on a single chip, with the ultimate objective of including the entire genome. At the same time, quicker analyses are being made possible by performing multiple analyses in parallel rather than in series to generate more information in lesstime. By using nanomaterials as sensing particles, chips could be reduced in size.

This would allow scientists to read thousands of molecules with the possibility of using cheaper equipment. Smaller, portable, cheaper and more precise integrated diagnostic kits for diseases such as systemic lupus or other multi-marker diseases may be made possible by using microfluidic and nanofluidic technologies. DNA arrays typically perform one type of analysis thousands of times. In the event that other experiments or processes are also required, micro and nanofluidic devices such as lab-on-a-chip, can integrate mixing, moving, incubation, separation, detection and data processing in a small portable device. At the micro and nanoscale, fluids move through pipes in laminar flow, as opposed to turbulent flow at the macro level.

This provides the opportunity for microfluidic devices to exploit certain physical behaviours. Two liquids can separately circulate through micro pipelines and valves without mixing each other. The combination of arrays and fluidic capabilities can greatly increase the speed and accuracy of various genetic testing. Within the drug discovery and diagnostic fields, microfluidic devices in the form of nanoarrays or "lab on a chip" technologies could allow the production of more efficient and disposable DNA and protein sequencers for drug discovery and diagnostic kits.

Imaging

Nano and micro technologies offer new imaging technologies that provide high quality images not possible with current devices, along with new methods of treatment. Malignant tumors are highly localized during the early stage of their development. If detected early, the tumors can often be surgically removed with high success. The longer a patient has a malignant tumor the more likely the cancer will spread to neighbouring lymph nodes and other anatomic structures. Aggressive surgery or chemotherapy, or very high doses of radiation may kill the cancer but at the same time severely injure normal tissue.

In many patients, particularly those with breast cancer, the cancer can spread to other parts of the body even after the original cancer tumor has been removed. Brain tumors have other challenges. Removing tumors and other cancerous cells in the brain through surgery can risk the loss of cognitive ability, hearing, speech, and other functions. Chemotherapy therapeutics can kill cancerous cells but because of their large physical size may be prevented from entering the brain by the blood-brain barrier. Nanotechnology can offer new solutions for the early detection of cancer and other diseases. Complementary technologies can also destroy cancer cells, which are too small to locate and remove surgically.

Nanoparticle Probes

Researchers at the University of Michigan are developing nanoprobes that can be used with magnetic resonance imaging. Nanoparticles with a magnetic core are attached to a cancer antibody that attracts cancer cells. The nanoparticles are also linked with a dye which is highly visible on an MRI. When these nanoprobes latch onto cancer cells they can be detected on the MRI.

The cancer cells can then be destroyed by laser or low dosage killing agents that attack only the diseased cells. In brain tumors, the presence of cancer can weaken the blood-brain barrier. Research is being conducted to have nanoprobes cross the weakened barrier to the tumor but not cross into healthy brain tissue. Another group at Washington University is using nanoparticles to attract to proteins emitted from newly forming capillaries that deliver blood to solid tumors. The nanoparticles circulate through the bloodstream and attach to blood vessels containing their complementary protein. Once attached, chemotherapy is released into the capillary membrane.

The nanoparticles traveling in the bloodstream would be able to locate additional cancer sites which may have spread to other parts of the body. Nanotechnology can also be used to improve the contrast agents for MRI's. Using fullerenes to encapsulate contrast agents such as gadolinium, Luna Innovations claims that traditional images can be enhanced by 50 times. If successful, this can result in the use of lower magnetic fields through smaller, cheaper and portable imaging equipment. Alternatively, smaller amounts of contrast agents could be used which would reduce the risk of allergic reaction in some patients.

Miniature Imaging Devices

Miniature wireless devices are being developed with nano and micro technologies for providing high quality images not possible with traditional devices. Given Imaging has developed a pill containing a miniature video system. When the pill is swallowed, it moves through the digestive system and takes pictures every few seconds. The entire digestive system can be assessed for tumors, bleeding, and diseases in areas not accessible with colonoscopies and endoscopies. Another company, MediRad is trying to develop a miniature x-ray device that can be inserted into the body.

They are attempting to make carbon nanotubes into a needle shape cathode. The cathode would generate electron emissions to create extremely small x-ray doses directly at a target area without damaging surrounding normal tissue.

4

Nanomedicine: The Medical Revolution

A NEW MEDICAL TECHNOLOGY

Living systems can usually heal and cure their own injuries, unless those injuries are severe enough to prevent the living system from functioning. Too often, we suffer injuries that are indeed this severe. Molecular nanotechnology is feasible. As we master the ability to design molecular machines that can continue to function when the living system around them has failed, those molecular machines can restore the function of the living system. They can support and sustain the processes of the living system until that living system can once again function on its own. Whether this is done by a temporary assist from respirocytes or by any of the myriad other techniques discussed in *Nanomedicine*, the underlying message is clear: life and health can be restored and sustained in the face of greater injury, greater damage, greater trauma, and greater dysfunction than has ever before been realised. This will usher in a new era of medicine – an era in which health and long life will be the usual state of affairs while sickness, debility and death will be the mercifully rare exceptions.

The future capabilities of nanomedicine give hope and inspiration to those of us who still have decades of life to look forward to, but some are not so fortunate. Many others who rightfully should live several decades more might find that chance cuts short their expected time. Heart attacks and cancer can strike us down even in the prime of our lives. They do not always wait their turn and politely arrive only when expected. How can today's dying patient take advantage of a future medical technology that is as yet only described in a handful of theoretical publications? How can we preserve the physical structure of our bodies well enough to permit that future medical technology to restore our health?

The extraordinary medical prospects ahead of us have renewed interest in a proposal made long ago: that the dying patient could be frozen, then stored at the temperature of liquid nitrogen for decades or even centuries until the necessary medical technology to restore health is developed. Called cryonics, this service is now available from several companies. Because final proof that this will work must wait until after we have developed a medical technology based on the foundation of a mature nanotechnology, the procedure is experimental. We cannot prove today that medical technology will be able to reverse freezing injury 100 years from now. But the patient dying today must choose whether to join the experimental group or the control group. The luxury of waiting for a definitive answer before choosing is simply not available. So the decision must be made today, on the basis of incomplete information. We already know what happens to the control group. The outcome for the experimental group has not yet been confirmed. But given the wonderful advances that we see coming, it seems likely that we should be able to reverse freezing injury — especially when that injury is minimized by the rapid introduction through the vascular system of cryoprotectants and other chemicals to cushion the tissues against further injury.

The development of nanomedicine depends on us: what we do and how rapidly we do it. Research is not done by a faceless "them," nor is it something that happens spontaneously and without any human intervention. It is done by and supported by people. Unless we decide to support and pursue this research, it won't happen. How long it takes to develop depends on us. We are not idle bystanders watching the world go by. We are a part of it. If we sit and wait for someone else to develop this technology, it will happen much more slowly. If we jump in and work to make it happen, it will happen sooner. And developing a life saving medical technology within our lifetimes seems like a very good idea – certainly better than the alternative.

Sometime in the next 30 years, mankind should acquire unprecedented ability to grasp, manipulate and modify individual molecules, and this ability may have profound implications. The term "molecular nanotechnology" has been coined to mean the technology of highly versatile and inexpensive molecular fabrication, molecular manipulation, and molecular-level manufacturing. It refers to the nanometer size range, the scale of atoms and molecules, but could be used to create macroscopic structures of precisely defined composition. The key concepts of molecular nanotechnology are that molecules can be machines and that molecular engineers can work with molecules to build desired equipment just as effectively as today's engineers now work with bulk materials. The difference between nanotechnologists and biotech-nologists is that the former do not restrict themselves to the biological limitations of the latter, and they are much more ambitious about the kinds of accomplishments that they want to achieve.

Organic chemists have long been able to synthesize complex molecular structures, including drugs, by devising ingenious reagents and by contriving reaction conditions in such a way as to minimize undesired side reactions and maximize yield. The results obtained depend on the statistics of uncontrolled molecular collisions in solution involving all possible molecular degrees of freedom. Statistically improbable reactions may be slow. More important, solution-based reactions generally lead to unwanted and potentially toxic byproducts. This imprecision of present day chemical synthesis and manufacturing processes is one reason that drug standards are needed. In contrast, molecular nanotechnology would directly and rapidly produce the desired chemical and only the desired chemical, giving yields of 100 Percent. This would be achieved by precisely positioning and bringing together the individual molecules involved in the reaction in such a way as to catalyze the reaction desired and only the reaction desired.

PATHWAYS TO MOLECULAR NANOTECHNOLOGY

Molecular nanotechnology has many precedents. Enzymes are natural molecular machines that adsorb individual reactant molecules from the surrounding solution and, as a result of precisely orienting them with respect to each other in a protected "nanoenvironment," catalyze reactions in a highly specific manner at very high speeds and under mild reaction conditions. This simple process in biological systems ultimately allows synthesis of structures as diverse as carbon dioxide and hair. In fact, living organisms are naturally-existing, fabulously complex systems of molecular nanotechnology. If nature can produce the biochemical capabilities of living cells by accident, molecular engineers should be able to accomplish comparable, but broader capabilities by design, guided in part by the examples provided by living systems.

Many industries already make use of enzymes to catalyze desired reactions one molecule at a time. Genetic engineers are producing pharmaceuticals by using naturally occurring enzymes to edit DNA, and the soft drink industry uses enzymes that have been modified to allow them to produce sugar at high rates near 100 degrees C without denaturation. The real promise for the future, however, lies in the development of fully artificial enzymes. Enzymes have already been designed, synthesized, and found to function as designed. Designed enzymes that are found not to function as intended can be modified as many times as necessary until they function as desired. Thus, whether from first principles or from enlightened trial and error, industrially useful artificial enzymes should be forthcoming. More than 10^{100} average-sized synthetic enzymes are possible with the use of nature's 20 amino acids, whereas probably less than 10^{14} enzymes are presently responsible for maintaining the entire biosphere.

But there is nothing to restrict artificial enzymes to only 20 amino acids. There are many useful kinds of chemistry that are not easily promoted by natural amino acids. One particularly versatile method for transcending the biological limits of proteinaceous catalysts has already been demonstrated. It is based on the fact that the genetic code specifies amino acids by the sequence of any one of four nucleic acid bases taken three at a time, with each three-letter base sequence known as a codon. There are 64 codons in all, but nature uses them to specify only 20 amino acids rather than 64. Recently, it has become possible to create an artificial transfer RNA that can add an unnatural amino acid to a growing polypeptide chain in response to one of the "unused" codons that exist naturally. This could be expanded to an unlimited number of unnatural amino acids, and these unnatural amino acids could contain totally nonbiological catalytic groups or even pre-made machine parts, such as structural support struts, molecular bearings, or the like. In a fully artificial system, the 20 natural amino acids might even be entirely dispensable. Furthermore, it is now possible to insert artificial bases into DNA and RNA, drastically augmenting the prospects for designing catalytically active RNA as well as proteins with unnatural functional groups. The potential for programming the creation of unprecedented chemical catalysts and other molecular tools useful for molecular engineering is thus virtually open-ended.

Yet a ribosomal pathway to molecular nanotechnology is far from the only viable approach. Bulk techniques are now creating useful building blocks for molecular nanotechnology, such as carbon closed-end tubes with walls that are one atom thick. Cryptands or cavitands are being created that pack the catalytic punch of enzymes into nonprotein-aceous structures far smaller than natural enzymes. At the same time, the ability of the scanning tunneling microscope not only to image, but also to manipulate individual atoms and molecules is being combined with the biological specificity of antibodies in an attempt to make all-purpose molecular synthesizers similar in concept to industrial robots that now make automobiles.

These present-day efforts are being supplemented by computational chemistry simulations of molecular bearings, molecular planetary gears, and molecular robot arms that would direct molecular factories.

A New Era of Medicine

An important premise of molecular nanotechnology is that the products of this technology should ultimately be readily affordable. This premise is based on the notion that an all-purpose molecular assembler should be able to make copies of itself. Thus, once the first programmable molecular assembler is made, it will not be long before there are as many such assemblers as the market needs.

Designs for computers with molecular data storage and processing elements predict data storage capacities of 1,000 megabytes per cubic micron and data-processing capabilities of 10^{10} operations per second for similar volumes. Translated into a sugar-cube-sized personal computer, this molecular computer would store 10^{20} bytes of information and process information at a speed of 10^{20} operations per second. This is 200 billion times the information storage capacity of today's top level personal computer and well over 100 million times the processing speed of today's fastest supercomputers.

Although it is difficult to project how many of the open-ended possibilities suggested by molecular nanotechnology will be realised by the year 2020, several innovations seem worth considering.

Quality control in the pharmaceutical industry today is necessary due to the imprecision of the manufacturing, purification, and packaging processes and the inability to monitor product quality on a continuous basis. When drugs are made by programmable molecular fabricators and when molecular sensing devices are available, continuous sensing of the near-flawless production process should be possible, both for internal use and for reporting purposes. Problems could be corrected instantly by replacing defective chemical synthesizers with backup copies and discarding the few molecules of mis-synthesized material before any has a chance to leave the factory.

Thus, the products reaching the public should conform to standards with virtually exact fidelity. It is also possible that multipurpose drug synthesizers could be on hand in most clinical chemistry and toxicology laboratories and would make on-site *de novo* synthesis an alternative to obtaining pre-synthesized standards from bodies such as the USP. The barriers to this route could be more political than technical, but the availability of rapid local synthesis could make a life-or-death difference in some cases, such as those involving acute poisoning. Alternatively, and more probably, instruments capable of identifying and quantitating drugs without calibration standards should be feasible.

"Smart" Pharmaceuticals

Today's drug is essentially a single molecule with an often sophisticated but always limited repertoire. Tomorrow's "smart pharmaceuticals" could be essentially programmable machines with a range of "sensory," "decision-making," and "effector" capabilities. They might avoid side effects and allergic reactions by coming in generic, biocompatible housings; becoming active only upon reaching their ultimate destinations; and attaining almost complete specificity of action. They might check for overdosage before becoming active, thus preventing accidental or intentional poisoning. They might have not one chemical action but several, processing targeted invading organisms or malignant cells through a series of chemical reactions that guarantee the death of the target. They might work in concert with three or four "sister" agents that together produce versatility unattainable by one agent alone.

Despite the vast increase in complexity over present-day drugs, such agents can be expected to be totally "pure" and predictable in their behaviour. Safety and efficacy may be inherent in the designs of such "drugs," in which case regulatory issues could be simplified rather than complicated.

Drug and Health-Related Information

When every intelligent person can have essentially unlimited data storage and data-processing capability at his or her fingertips, drug information will change dramatically.

Marked advances in diagnostic agents could have far-reaching consequences. The year 2020 will occur approximately 15 years after the complete human genome has been obtained. By 2020, great advances in understanding biochemical individuality, which is so important for side effects and proper dose adjustment, will have been made. This may make it possible for the physician to read critical aspects of a patient's phenotype from a noninvasively obtained cell sample and to inject "smart" diagnostic agents that can be recovered in a drop of saliva some time later and read out to reveal signs of previously undetected, impending disease processes. The opportunities for precise tailoring of individual treatment and for preventive medicine, with all the cost savings implied by both, would be revolutionary.

"Drug information" could come to include the complete matrix of appropriate pharmacologic responses to the newly-available individuality data. Eventually, this will become so complex and extensive that the physician will be utterly dependent on assistance provided by computer "expert systems" in making decisions. In the long run, particularly if ingestible diagnostic agents and home readout systems become available, the need for physician participation in pharmacological therapy may vanish altogether for those patients with the right equipment available. A major incentive for this decentralization of medical care will be ethical concerns over patient privacy and the use of genotypic information for unauthorized purposes: these are issues of "drug information" that continue to redefine the term. Relevant compendia should be available on line to any interested party, particularly those whose software allows them to interface the data with a medical "expert system."

Information about "drugs" per se will also change dramatically. Drug documentation will come to resemble today's protein structural databases because of the potentially great complexity of "pharmaceutical agents" that differ from medical devices only in that they are injectable, and/or invisible to the naked eye. Indeed, the line between pharmaceuticals and medical devices will become fuzzier and fuzzier. On the other hand, the need to document contra-indications, side effects, and complications of use should be considerably reduced.

The world of health care technology will be substantially different in 2020 from that of today. The Pharmacopeial Convention will likely be a much different organization on the occasion of its bicentennial anniversary. The need for pharmaceutical scrutiny by medical practitioners could be even more intense, however, as all aspects of pharmaceutical medicine and the pharmaceutical industry itself continue to change dramatically and at great speed. It can be hoped that molecular nanotechnology, in addition to helping to create these dramatic changes, will also be one of the technologies that help the practitioners of 2020 stay abreast of and manage these developments.

CURRENT TECHNOLOGY

Currently the area of nanomedicine showing the greatest potential is the use of tiny polymeric drug carriers. Such materials can now be made in the form of spheres with diameters ranging from about 50 billionths of a meter. This can facilitate the transport of drugs to some of the smallest capillaries in the body. These nanospheres are designed to travel to specific sites within the body, release their payload of drug molecules and then degrade. Their degradation by-products are non-toxic and will ultimately be excreted from the body. Developments in polymer science have had a great influence on drug delivery. It is now possible to synthesise a wide variety of biocompatible biodegradable polymers that will release entrained drugs at a rate determined by the chemistry and physical form of the polymer. Drug delivery research is now at a very important stage in its development.

In recent years there has been revolutionary progress made in the field of healthcare. The knowledge now exists to cure many afflictions that were once thought incurable through the use of novel therapeutics. Many of these therapeutic methods cannot currently be used or are not being used to their full potential because they cannot be delivered into the body with the required control. An example of this is chemotherapy for the treatment of cancer. This involves saturating the body with toxic drugs, which can often result in harmful side effects such as reduced immune response to infection. By directing the drug molecules specifically to the site of a tumour a reduced quantity of drug would be required thus reducing the toxic effect on the body.

Advances in biotechnology now make it possible to control disease at the genetic level. These developments have lead to the creation of several new fields in the area of biopharmaceutical science. These new therapies are based on the interaction of nucleic acid drugs with the genetic material of specific cells within the body. The Achilles' heel of these new biopharmaceuticals is that it is difficult to direct them to the specific site required for action. There is also another disadvantage to these therapies in that they are easily broken down by conditions within the body. Incorporation of such drugs into delivery devices may be the way forward.

So how are such nanoparticulate devices produced? The answer is a complicated one that involves the use of molecules that self-assemble themselves during a chemical process. This procedure, which is known as 'emulsion polymerisation'. This involves the use of self-assembled spheres as a molecular 'mould' into which reactive monomer diffuses and polymerises to form solid spheres.

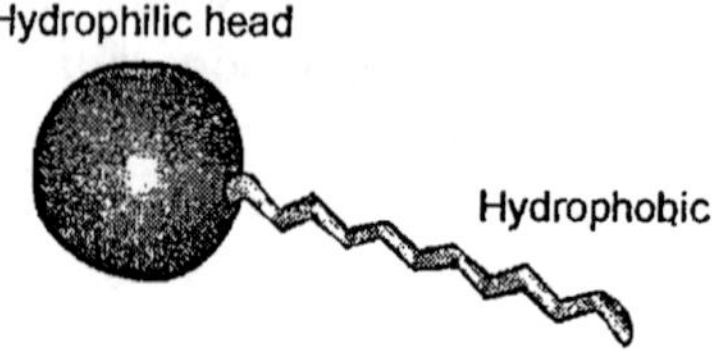

Fig.: Schematic Representation of a Surfactant Molecule

Molecules known as surfactants perform this bizarre self-assembly by virtue of their unusual structure. The process has been likened to putting transistors, resistors and diodes into a bag, shaking it around and pulling out a fully functional radio. The mechanism that causes surfactant molecules to self-assemble is a result of their amphiphillic nature. This means that the molecules are composed of two or more parts with each part being soluble in a different medium. Generally, a surfactant may have a hydrophilic head and a hydrophobic tail.

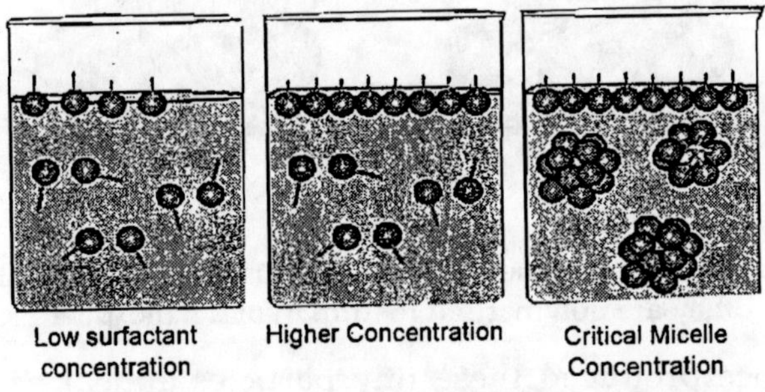

Fig.: The self assembly mechanism of micelle formation with increasing concentration of surfactant

Once the concentration of these molecules in an aqueous solution reaches a critical value they will arrange themselves so as to minimise the interaction of the hydrophobic tail with the water. In order to accomplish this the hydrophobic tails align themselves along side each other that, in some cases lead to the formation of a spherical structure called a micelle. The size of these micelles is dependent on the length of the surfactant molecule and range from 10 - 100 billionths of a meter in diameter. Once there are millions of micelles spinning around in the aqueous medium a reactive monomer is introduced which is also hydrophobic. Monomer molecules want to minimise their interaction with the water and for this reason they travel to the interior of the micelles where the are stable. The micelles will then swell to accommodate the monomer. The final part of the preparation of the nanoparticles involves the addition of initiator molecules that trigger a chain reaction within the micelles core leading to a polymerisation. Once the core of the micelles have been converted into a solid polymer sphere of approximately 200 - 1000 billionths of a meter in diameter the spheres may be removed from the emulsion and are ready for use.

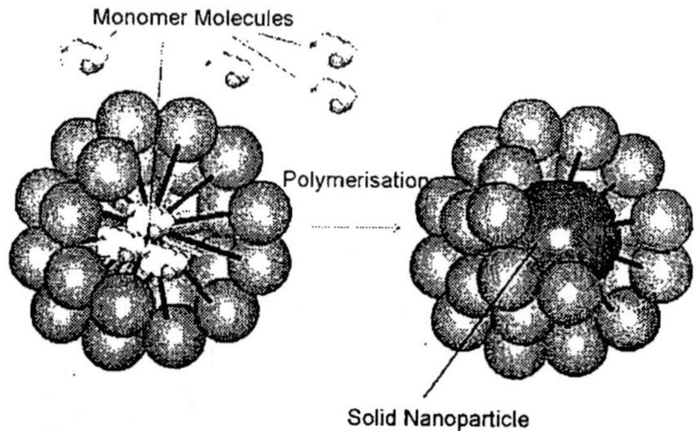

Fig.: Once monomer molecules diffuses into the micelle they are polymerised to form a solid nanosphere

Once produced these tiny particles appear as a fine powder that can have drug molecules either attached to their surface or absorbed into their core. At this stage the nanoparticles are solid spheres composed of polymer chains of a fixed length. This chain length is described by the polymer's molecular weight. For a specific polymer, an example being poly(n-butyl cyanoacrylate), the lower the molecular weight the faster it will degrade within the body and thus the faster the entrained drug is released. The specific site within the body to which the nanoparticles will travel is controlled by the surface chemistry of the nanoparticle. Molecules can be tethered to the particles' surface, which will cause selective uptake, by various organs within the body.

FUTURE RESEARCH

Advances in analytical techniques have improved our ability to produce and image nanostructures. Scanning Electron Microscopy uses an electron beam to image nanoparticles. They cannot be seen using standard optical microscopy because their diameter is smaller than the wavelength of visible light. SEM however has its disadvantages in that the powerful electron beam used can often vapourise the particles. A new technique is now becoming available that generates an image by passing a stylus back and forth over the surface of the particles. This stylus is much like the one used on a record player but is a few thousand times smaller. The technique is called Atomic Force Microscopy and its magnifying capability is such that it has been compared with the ability to count the grains of sand on a beach on earth, from the moon! At the moment there is a great deal of research being carried out in the area of nanotechnology. Materials that change their physical characteristics following exposure to stimuli such as light, heat and changes in pH are being developed. It has been proposed that such materials will lead to the first nano-scale robot arms and similar devices. By combining technologies it may be possible to create 'smart' nanoparticles that will be the next step towards true cellular level medicine.

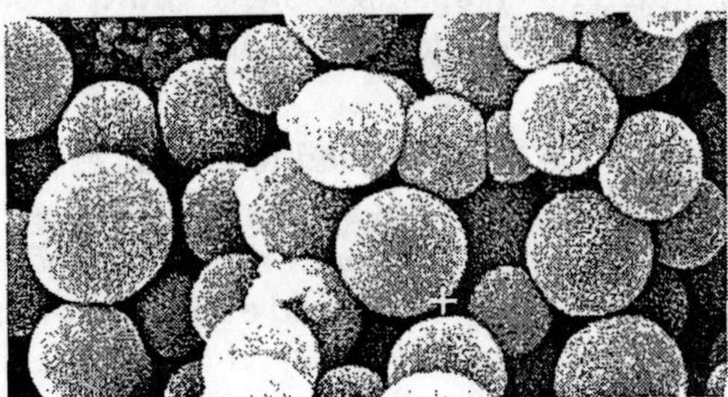

Fig.: Scanning Electron Microscope image of nanoparticles

NANOMEDICINE: THE BIGGER PICTURE

Future techniques in medical diagnosis and treatment have often been the subject of science fiction literature and cinema. The approaches are surprisingly consistent. The diagnosis of illness and its treatment is generally non-invasive and mostly painless. In 'The Six Million Dollar Man', an individual who has undergone serious accidental damage is given replacement organs and tissues that function as well, or even better than the originals. What was once the stuff of science fiction is now closer to becoming a reality.

Nature operates at the nanoscale, and today we are acquiring an increasingly profound understanding of natural processes at this scale, enabled by a new generation of scientific instruments. From this knowledge, we are able to design devices that can either directly interact with, or influence, the behaviour of living cells. Nanotechnology has a trump card to play when applied to medicine. At the nanometre scale, materials often exhibit surprisingly different physical, chemical, and biological properties when compared with the same material in bulk form. The properties of nanoparticles, such as increased chemical activity and the ability to cross tissue barriers, are leading to new drug targeting and delivery techniques. In the future, it is within the realms of possibility that a nanoparticle may be designed to search for, find and destroy a single diseased cell, taking us even closer to realizing the ultimate goal of disease prevention. Nanotechnology is also making possible other techniques, such as the stimulation of the body's own mechanisms to successfully repair diseased or damaged tissues, replacing the need for transplants and artificial organs.

In the forseeable future, allied to advances in information and communication technologies, nanotechnology as applied to medicine, will lead to advances in remote monitoring and care, where a patient may be treated at home - a less expensive option, and one that is more conducive to a successful medical outcome than treatment in a surgery or hospital. Continued research into disease processes at the molecular level is essential for the development of nanomedicine, and involves teams of scientists from across 'conventional' disciplines, such as physics, chemistry, surgery and mathematics, as well as those from the 'new' fields of genomics, proteomics, metabolomics, pharmacokinetic modelling and even microscope design.

Challenges exist in training and managing these multidisciplinary teams and partnerships, and importantly, in finding new solutions to the intellectual property issues that presently hinder the speedy commercialization of new knowledge. Research challenges also relate to understanding and modelling the toxicity of engineered nanoparticles, and requires that toxicologists work alongside medical scientists wherever nanoparticles are involved in drug targeting and contrast agent development. There are technological challenges, too, in the areas of molecular manufacturing, quality assurance and the eventual programmability of nanodevices.

Nanomedicine will also bring its own set of new legal and ethical challenges to be resolved. Cost benefit analyses are also critical. Where money can be best invested for the greatest social and economic benefit needs to be carefully decided. Should investment be made in prevention or treatment; in diseases of the poor or the rich; in long term hospitalization or costly drug development and deployment? Any strategy for nanomedicine must also be influenced by the facts that resources are finite and demands are great. Within these constraints the health needs of an ageing population need to be managed, cures found for major lifestyle 'killer' diseases, such as cancer, and for the diseases of the less-developed world, such as HIV / aids, malaria and tuberculosis. In conclusion, nanoscience and nanotechnology are leading to extraordinary new breakthroughs in medicine that were once the stuff of dreams. EuroNanoForum aims to provide a small but representative snapshot of these exciting opportunities.

TISSUE ENGINEERING, NANOSCAFFOLDS, AND INTERFACES

Science is moving from transplanted organs to implanting of substitute or artificial organs to stimulating the body to do its own repairs. Tissue regeneration is about enabling the body to ultimately regenerate its own diseased or failed organs. The market worldwide for tissue engineered products already worth $18bn, and many hundreds of thousands of people are awaiting transplants worldwide. This has led to an illegal market in donor organs from the poor to the rich, so the development of effective regeneration techniques offers a multitude of benefits.

Tissue Regeneration Work

Instead of being surgically repaired, transplanted, or even fixed using prosthetics, tissue or organ failure could be solved by implanting natural tissue and organ mimics which can be fully functional from the start, or grow into the required functionality. Nanotechnology helps to recruit the body's natural healthy cells to promote regeneration of tissue on or around a damaged area. It can also act as a 'scaffold' to provide a framework for developing tissues to latch onto and penetrate. This application of nanotechnology will result in reconstructed tissue and wound treatments that are superior, longer lasting and more acceptable to everyone involved, most notably the patient.

The beauty of the process is that it is arguably the most natural way of healing, as it is the body's own healthy cells which are regenerating - as they were meant to do - albeit with a small push to get started. Tissue regeneration would be almost unthinkable without development in nanoscience and nanotechnology. It underpins the design of scaffolds on which the cells are grown, which are composed of special nanocomposite materials hat contains cell growth stimulants, and have nanoscale, cell-friendly surface topographies. There is even the potential for whole organs to be grown to replace those that have failed through disease or old age.Tissue regeneration offers a revolution in healthcare, with huge benefits for doctors who will be using the technology, to the government who will save money, and most importantly to the patients who are the ultimate beneficiaries.

DRUG DELIVERY AND PHARMACEUTICAL DEVELOPMENT

A long standing issue for pharmaceutical companies is to deliver the right amount of drug to the site of disease. The inability to achieve this has means that drugs need to be administered in excessively high doses, increasing the likelihood that patients may suffer from toxic side effects.

Nanotechnology enables drugs to be targeted more effectively to the site of the disease, and activated 'on arrival'.

Drug can be prepared in such a way that they can cross epithelial barriers, and differentiate between healthy cells and diseased ones. Very small drug particles (nanoparticles) are coated with special molecules and travel round the body until they reach the disease site. Once there, the molecules 'decorating' the drug particles recognise the disease site, bind to it, and the drug is then activated. This targeting means the drug will work efficiently and more effectively, and side-effects dramatically decrease. Chemotherapy in cancer usually means healthy cells are also killed off during the attempt to kill the cancerous cells. Targeted drug delivery circumvents this, and offers the potential for a quicker recovery.

Nanotechnology has other benefits for therapy. In nanoparticulate form, drugs are less prone to degradation, meaning they will have a longer shelf-life. Nanoparticluate drugs can be coated to delay their activation, resulting in the drug remaining in the body longer, reducing the frequency of dosage required. Some drugs may be linked to particles which fluoresce when they reach diseased cells. This is a special technique known as find, fight and follow, and is leading to early detection and treatment. The area of disease can be identified by imaging the fluoresecent particles, which then deliver their payload of drugs, and the effectiveness of the drugs can be then monitored. In essence, the application of nanotechnology to drug targeting and delivery offers more effective treatments, lower toxicity and side-effects and reduced cost from reduced dosage. Drug delivery and targeting are areas which are already offering real patient benefits.

CELL STRUCTURE AND FUNCTION

Humans are essentially a collection of cells! The better we understand how cells work, the more likely it is we can create effective cures, understand the cause of disease better and produce better drugs. Nanotechnology is enabling this understanding. This session is concerned with understanding of cell structure and function and how a cell manifests disease, and the new tools that are being developed to identify early signs of disease at the cellular level, monitor of healthy and diseased cells and the progress of therapy, and measure the effects of different drugs.

Once we know the signals that a cell gives when it is diseased we can look for that signal. Handin-hand with the growing understanding of cell function, is the technology for measuring and identifying changes that indicate disease. The development of 'lab-on-a-chip' is one such technology which has many applications in disease identification.

These include enabling cells to be studied independently of the body, the speedy screening patients for infection, and the analysis of the hereditary propensity to disease. A lab-on-a-chip is a miniature laboratory that uses techniques developed from the electronics industry, that offers speed, sensitivity, accuracy and portability, compared with more traditional techniques, which are slow, expensive resource intensive and may be inaccurate as samples deteriorate with time. In the future, this technology is leading to patient-centred medicine, with patients conducting tests in at their own homes, and downloading the results via their computer to a medical monitoring centre.

Lab-on-a-chip technology also has important applications in being able to quickly screen a large population for disease or potentially killer infections, such as avian flu. Lab-on-a-chip techniques and a knowledge of the human genome is leading to pre-symptomatic disease treatment through identify the individual genetic propensities say to breast or colon cancer or heart disease, so appropriate preventative action can be taken.

This technology is also important for the developing world specifically in some African countries where disease is at epidemic rates. In areas where very little money, or even healthcare is available, the provision of cheap, fast-acting LOCs could offer early identification of disease, and enable more appropriate effective and cheaper treatment options. The cost effectiveness of what is basically a tiny microchip is another selling point; it can be mass produced incredibly cheaply.

THE PROMISE OF NANOMEDICINE

Nanomedicine offers a huge spectrum of benefits for the better treatment of disease - from early diagnosis to better therapies to eventually preventative medicine; from stimulating the body to regenerate its own diseased tissues to the targeting and activating drugs at the site of diseae. It also offers the possibility of developing better drugs faster, that have fewer side effects and are customised to the patient's own genome.

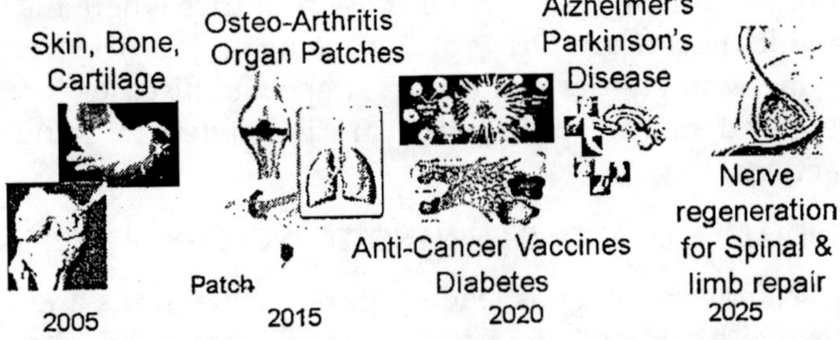

Fig.: Application of Nanotechnology in Mdeicine

Three particular areas where nanotechnology will have a major impact in relation to cancer; in relation to reducing animal testing, and in the commercial impact of new nano-based therapies:

Cancer treatment is the past has suffered from a lack of understanding the mechanisms of the disease at the cellular level, late diagnosis, and treatments based on attacking the whole body rather than the disease site. Nanotechnology is offering a totally new approach; from a better understanding of the root causes of disease, to quicker diagnosis and science-base treatments. Today, to quote Dr Mauro Ferrari ' we are seeing the tip of the iceberg' in terms of the potential nanotechnology offers for cancer treatment.

Another area where we are using a hammer to crack a nut is in the area of animal experiments for the validation of drugs. This has obviously produced some benefits in the past, but these are relatively small in relation to the large body of research as many apparently good results could not be adequately replicated in humans, or produced unexpected side effects. With a better knowledge of individual responses to chemicals, nanotechnology is leading to a refinement of techniques for drug discovery that are accurate and meaningful for the individual patient, based on the use of living cells.

Nanotechnology is leading to better treatments based on our knowledge of proteomics and genomics, this is leading to a winwin situation for drug companies and patients as well as huge economic benefits for those countries where the State provides healthcare. Drugs are becoming more efficient as they are increasingly designed to treat specific diseases, patients are cured more rapidly, less care is needed to counteract reactions.

GENERAL APPLICATIONS OF NANOTECHNOLOGY IN MEDICINE

Nanotechnology is leading to faster diagnosis and more effective treatment, the early signs of disease, and its exact location. One such application uses quantum dots to locate the disease site.

Quantum dots are semiconductor nanoparticles which can glow very brightly under certain criteria. The quantum dots can be attached to chemicals which link to diseased cells. When the link is made, the quantum dots glow. This allows the sites of disease to be accurately pinpointed, and also malignant and non-malignant tumours can be differentiated. Quantum dots may help in identifying cells which may develop problems in the future. By recognising certain indicators that cannot be detected using conventional technologies, quantum dots can call attention to potential disease sites, leading to the ideal of disease prevention. Improved drug targeting and delivery is another highly promising area of nanomedicine, which is actually happening now.

Coated drugs in nanoparticulate form can be delivered directly to the diseased cells. This means smaller or less frequent doses are required, and toxicity is reduced as the targeted drug works more effectively. Therefore nanomedicine is leading to the reduction of undesirable and costly side-effects of drugs, as the customisation and targeting of treatment would allow for more control over where the drug goes and what it affects in the body. Nanotechnology is set to revolutionise healthcare through moving us towards preventative, or pre-emptive medicine, with many health and cost benefits, both to the State and the patient. Impacts are especially envisioned for the ageing and other hitherto disadvantaged populations.

CONVERGING TECHNOLOGIES FOR MEDICINE AND HEALTHCARE

Limits and barriers on medical research are being rapidly broken down by the convergence of many disciplines, such as nanotechnology, molecular biology with physics, information technology and even, at last, the humanities! Together, this is resulting in rapid advances in new techniques, processes and tools which are completely changing the way we have approached healthcare in the past.

It is not only researchers from different disciplines that are working together. Networks like Nano2Life are bringing private firms, scientists and research institutes together to solve the problems of treating disease. Strong partnerships are being formed that are providing a range of skills across the board, generating disruptive technologies and completely innovative solutions to old problems. These could include non-invasive diagnostics, individual-specific drugs, identifying disease at the cellular level, faster validation of new drugs, and stimulating the patient's own body to regenerate diseased or failed organs.

This session will examine convergence and its benefits on a European level. It is case studybased, describing networks and their component organizations, and the benefits that have accrued to industry from EU-supported partnerships and initiatives that connect complementary skills from a wide range of disciplines.

NANOTECHNOLOGY FOR CONGENITAL AND DEGENERATIVE DISEASES

Nanotechnology is expected to play an increasing role in the battle against congenital and degenerative diseases. As we live longer, many more diseases become manifest over a larger proportion of the population, such as Alzheimer's and Parkinson's. We are also experiencing an epidemic of lifestyle diseases, from diabetes to cancer to heart and vascular disease.

Firstly, early diagnosis is critical to many of these diseases, which is enabled by the miniaturization, user-friendliness, cheapness and wide availability of diagnostic techniques to such an extent that patients can monitor their own symptoms. Example include self-testing for diabetes and cholesterol levels. The spectrum of possibilities will expand dramatically; test results can be transmitted to a central health bureau, and therapies prescribed remotely. Working at the nanoscale has produced new hope for disease of the brain. If the therapy is drug-based, drugs in nanoparticulate form can now cross the blood-brain barrier. If the therapy requires regeneration of diseased tissue, new research is enabling the growth of the patient's own cells on nano-inspired scaffolds that can be then implanted into the brain.

Other areas where Nanotechnology offers hope is in the delivery of drugs in a way that emulated the body's own systems. In diabetes, insulin is required by the body at irregular intervals. Using techniques derived from the electronics industry, this need can be monitored using a small, silicon chip-based implant, and insulin delivered only as required. Electronics combined with nanotechnology is also the basis of better retinal and cochlear implants. Today, new retinal implants have been developed that enable people previously classified as blind to be reclassified as sighted. Present cochlear implants are crude, difficult to insert, awkward to wear and give poor performance, Future implants will use nanocomposite materials, be small and easy to insert and be cheap to produce. This also has very promising implications for the poor.

Nanoimaging and Diagnos

Medical imaging has advanced from a marginal role in healthcare to becoming an essential tool of diagnostics over the last 25 years. Molecular imaging and image-guided therapy is now a basic tool for monitoring disease and in developing almost all the applications of in-vivo nanomedicine. Originally, imaging techniques could only detect changes in the appearance of tissues when the symptoms were relatively advanced.

Later, contrast agents were introduced to more easily identify and map the locus of disease.

Today, through the application of nanotechnology, both imaging tools and marker/contrast agents are being dramatically refined towards the end goals of detecting disease as early as possible, eventually at the level of a single cell, and monitoring the effectiveness of therapy. The convergence of nanotechnology and medical imaging opens the doors to a revolution in molecular imaging, leading eventually to the detection of a single molecule or a single cell in a complex biological environment.

Engaging The Community

Nanomedicine offer many benefits for healthcare in the future. It is a complex subject, and additionally suffers from many misconceptions about the potential of Nanotechnology for medicine, arising out of a confusion between science fiction and fact, even amongst highly educated people. Considering governments and institutions the world over are spending many billions on nanotechnology research and development, a small amount of that funding would be beneficially channelled to wider public awareness and engagement.

As with any new technology, there are risks as well as benefits. Too frequently the downsides of an innovation are publicised, out of proportion to the benefits. A survey carried out last year by researchers at the University of North Carolina established that the more people knew about nanotechnology, the more they thought the benefits would greatly outweigh the risks. The opposite was also true. This is a key point which indicates a wellinformed public is likely to embrace nanotechnology and, whilst remaining wary of some of the risks, if they see it as bringing a major benefit to their lives. To inform the public, so they can make reasoned and reasonable decisions, and what needs to be put in place to provide reassurance that potential risks are receiving the right level of attention and response.

ADDRESSING DISEASES OF THE DEVELOPING WORLD

According to a study by the Canadian Programme on Genomics and Global Health at the University of Toronto Joint Centre for Bioethics - a leading international medical ethics think-tank - nanotechnology may provide the means to help developing countries across as spectrum of areas including improvement in water quality, reduction in environmental and early diagnosis and treatment and even prevention of killer diseases such as malaria, tuberculosis and HIV/AIDS.

Benefits of Nanotechnology

Diagnosis of disease.: About a quarter of all Africans are infected with AIDS. It kills millions each year and 95 Percent of all new cases of AIDS occur in developing countries. Nano-based devices are leading to easy and fast analysis of a range of diseases that can be undertaken cheaply and easily.

Treatment of Disease.: Nanomedicine also has the potential to ensure longer shelf life for drugs, easier administration, and lower dosage requirements. New high throughput screening techniques are leading to cheaper drug discovery; and cell based assays should lead to faster introduction of these drugs into the population.

Sources of Disease. Nanotechnology-enabled systems have application for water purification through membrane filters, which would have the knock-on effect of reducing infection.

Other: Apart from benefiting medicine directly, nanotechnology is also offering access to cheap renewable energy using new polymer based solar power collectors that are lightweight, cheap and efficient.

Needs of the Ageing Population

Nanotechnology could finally mean that growing old does not necessarily mean loss of faculties and a rapid reduction in quality of life. Nanotechnology is supporting the rehabilitation of the infirm and elderly through the development of intelligent learning prosthetic devices. These include retinal and cochlear implants. Through the development of nanocomposites, nanoelectronics, nanosensors, nanotransducers and biomaterials, the next generation of implants will be more effective, body friendly and less traumatic to insert.

They will reinvigorate the faculties of patients whose hearing or sight has been affected by accident, disease or age. Organ replacement is increasingly common across the generations. Transplantation is fraught with dangers such as rejection and side effects from drugs. As discussed under Tissue Engineering, research is leading to the growth of new organs from a patient's own tissues. The ultimate objective is to produce innovative scaffold materials which canm be seede with the patient's own cells, which, when implanted will regenerate bone, cartilage and skin tissues in the most natural way possible.

Medical textiles is an exciting field of innovation. 'Smart clothes'are being developed which will regulate temperature and even monitor the health of the wearer. This remarkable new technology poses even more questions regarding ethics – the ethics of who should benefit, if there is a choice to be made, where the line should be drawn, if any, at enhancing human performance, and whether new technology will lead to a further widening of the gap between rich and poor, advantaged and disadvantaged.

IMPLANTS AND DEVICES

Nanotechnology, new materials and nanoelectronics are the basis of novel retinal and cochlear implants. New retinal implants have been developed that enable people previously classified as blind to be reclassified as sighted. Present cochlear implants are crude, difficult to insert, awkward to wear and give poor performance, Future implants will use nanocomposite materials, be small and easy to insert and be cheap to produce. This also has very promising implications for the poor.

Over $30bn is spent on implants yearly (hip, knee, dental etc.). Nanotechnology can help bring about better implants. Proteins at the body-implant interface play a decisive role in the acceptance or rejection of the implant. Using nanotechnology we can produce 'bodyfriendly' coatings and also create topographies that cells like, to increase the likelihood of acceptance. These biocompatible coatings are even being used in devices like hearing aids,reducing the likelihood of irritation or skin reaction. Around 1,000,000 hip and knee-joint replacements are carried out in the EU each year. However, the lifespan of the implants is only around 10 years, shorter if the patient is particularly active or overweight. This poses both a quality of life problem, and a cost effectiveness issue. Revisionary surgery adds about €520m a year to the EU's medical costs.

Fig.: A ceramic implant

Nanocomposites are also being developed that form ceramic implants which are stronger and longer-lasting, with potential life-spans of more than 30 years.

NANOSENSORS AND DIAGNOSTICS

In nanodiagnostics, the ultimate goal is to identify disease at the earliest stage possible, ideally at the level of a single cell. When testing for disease, nanotechnology offers better sensitivity, specificity and reliability. Several steps from sample preparation to detection can be integrated into a single, tiny device. An important application is in cheap, reliable devices, based on nanotechnology, that can be used at the "point of care" in clinics by non-expert technicians, and eventually by individuals themselves. Advances in in-vivo diagnostics will rely mainly on imaging. In molecular imaging, the goal is to create sensitive, reliable agents that can detect, deliver and monitor therapy. This is the "find, fight and follow" concept, sometimes also known as *theranostics*.

The likely diseased tissue is firstly imaged, using target-specific contrast agents. These contrast agents can be combined with a drug; and the results of therapy can then be monitored, again using imaging techniques. The advances in diagnostics have presented major opportunities for healthcare, including a plethora of new personalised treatments and early diagnosis for individuals suffering from a wide variety of illnesses such as cancer, diabetes, and heart disease.

Nanotechnology has even made a tiny 'nano needle' possible which can provide important information on disease at the level of a single cell. Other types of nanosensors include cantilevers which are sensitive, simple and cheap and able to accurately detect tiny biochemical changes.

5

Medical Technology Trends

GENOMICS

By 2015, biotechnology will likely continue to improve and apply its ability to profile, copy, and manipulate the genetic basis of both plants and animal organisms, opening wide opportunities and implications for understanding existing organisms and engineering organisms with new properties. Research is even under way to create new free-living organisms, initially microbes with a minimal genome.

Genetic Profiling and DNA Analysis

DNA analysis machines and chip-based systems will likely accelerate the proliferation of genetic analysis capabilities, improve drug search, and enable biological sensors.

The genomes of plants ranging from important food crops such as rice and corn to production plants such as pulp trees and animals ranging from bacteria such as *E. coli*, through insects and mammals will likely continue to be decoded and profiled. To the extent that genes dictate function and behaviour, such extensive genetic profiling could provide an ability to better diagnose human health problems, design drugs tailored for individual problems and system reactions, better predict disease predispositions, and track disease movement and development across global populations, ethnic groups, and other genetic pools. Note that a link between genes and function is generally accepted, but other factors such as the environment and phenotype play important modifying roles. Gene therapies will likely continue to be developed, although they may not mature by 2015.

Genetic profiling could also have a significant effect on security, policing, and law. DNA identification may complement existing biometric technologies e.g., retina and fingerprint identification for granting access to secure systems e.g., computers, secured areas, or weapons, identifying criminals through DNA left at crime scenes, and authenticating items such as fine art. Genetic identification will likely become more commonplace tools in kidnapping, paternity, and fraud cases. Biosensors some genetically engineered may also aid in detecting biological warfare threats, improving food and water quality testing, continuous health monitoring, and medical laboratory analyses. Such capabilities could fundamentally change the way health services are rendered by greatly improving disease diagnosis, understanding predispositions, and improving monitoring capabilities.

Such profiling may be limited by technical difficulties in decoding some genomic segments and in understanding the implications of the genetic code. Our current technology can decode nearly all of the entire human gene sequence, but errors are still an issue, since Herculean efforts are required to decode the small amount of remaining sequences. More important, although there is a strong connection between an organism's function and its genotype, we still have large gaps in understanding the intermediate steps in copying, transduction, isomer modulation, activation, immediate function, and this function's effect on larger systems in the organism. Proteomics the study of protein function and genes is the next big technological push after genomic decoding. Progress may likely rely on advances in bioinformatics, genetic code combination and sequencing akin to hierarchical programming in computer languages, and other related information technologies.

Despite current optimism, a number of technical issues and hurdles could moderate genomics progress by 2015. Incomplete understanding of sequence coding, transduction, isomer modulation, activation, and resulting functions could form technological barriers to wide engineering successes. Extensive rights to own genetic codes may slow research and ultimately the benefits of the decoding. At the other extreme, the inability to secure patents from sequencing efforts may reduce commercial funding and thus slow research and resulting benefits.

In addition, investments in biotechnology have been cyclic in the past. As a result, advancements in research and development may come in surges, especially in areas where the time to market is long.

CLONING

Artificially producing genetically identical organisms through cloning will likely be significant for engineered crops, livestock, and research animals.

Cloning may become the dominant mechanism for rapidly bringing engineered traits to market, for continued maintenance of these traits, and for producing identical organisms for research and production. Research will likely continue on human cloning in unregulated parts of the world with possible success by 2015, but ethical and health concerns will likely limit wide-scale cloning of humans in regulated parts of the world. Individuals or even some states may also engage in human or animal cloning, but it is unclear what they may gain through such efforts. Cloning, especially human cloning, has already generated significant controversies across the globe. Concerns include moral issues, the potential for errors and medical deficiencies of clones, questions of the ownership of good genes and genomes, and eugenics. Although some attempts at human cloning are possible by 2015, legal restrictions and public opinion may limit their extent. Fringe groups, however, may attempt human cloning in advance of legislative restrictions or may attempt cloning in unregulated countries

Although expert opinions vary regarding the current feasibility of human cloning, at least some technical hurdles for human cloning will likely need to be addressed for safe, wide-scale use. "Attempts to clone mammals from single somatic cells are plagued by high frequencies of developmental abnormalities and lethality". Even cloned plant populations exhibit "substantial developmental and morphological irregularities". Research will need to address these abnormalities or at the very least mitigate their repercussions. Some believe, however, that human cloning may be accomplished soon if the research organization accepts the high lethality rate for the embryo and the potential generation of developmental abnormalities.

GENETICALLY MODIFIED ORGANISMS

Beyond profiling genetic codes and cloning exact copies of organisms and microorganisms, biotechnologists can also manipulate the genetic code of plants and animals and will likely continue efforts to engineer certain properties into life forms for various reasons. Traditional techniques for genetic manipulation such as cross-pollination, selective breeding, and irradiation will likely continue to be extended by direct insertion, deletion, and modification of genes through laboratory techniques. Targets include food crops, production plants, insects, and animals.

Desirable properties could be genetically imparted to genetically engineered foods, potentially producing: improved taste; ultra-lean meats with reduced "bad" fats, salts, and chemicals; disease resistance; and artificially introduced nutrients. Genetically modified organisms can potentially be engineered to improve their physical robustness, extend field and shelf life, tolerate herbicides, grow faster, or grow in previously unproductive environments.

Beyond systemic disease resistance, *in vivo* pesticide production has already been demonstrated in corn and could have a significant effect on pesticide production, application, regulation, and control with targeted release. Likewise, organisms could be engineered to produce or deliver drugs for human disease control. Cow mammary glands might be engineered to produce pharmaceuticals and therapeutic organic compounds; other organisms could be engineered to produce or deliver therapeutics. If accepted by the population, such improved production and delivery mechanisms could extend the global production and availability of these therapeutics while providing easy oral delivery.

In addition to food production, plants may be engineered to improve growth, change their constitution, or artificially produce new products. Trees, will likely be engineered to optimize their growth and tailor their structure for particular applications such as lumber, wood pulp for paper, fruiting, or carbon sequestering while reducing waste byproducts. Plants might be engineered to produce bio-polymers for engineering applications with lower pollution and without using oil reserves. Bio-fuel plants could be tailored to minimize polluting components while producing additives needed by the consuming equipment.

Genetic engineering of microorganisms has long been accepted and used. *E. coli* has been used for mass production of insulin. Engineering of bacterial properties into plants and animals for disease resistance will likely occur.

Other animal manipulations could include modification of insects to impart desired behaviours, provide tagging including GMO tagging, or prevent physical uptake properties to control pests in specific environments to improve agriculture and disease control.

Research on modifying human genes has already begun and will likely continue in a search for solutions to genetically based diseases. Although slowed by recent difficulties, gene therapy research will likely continue its search for useful mechanisms to address genetic deficiencies or for modulating physical processes such as beneficial protein production or control mechanisms for cancer. Advances in genetic profiling may improve our understanding and selection of therapy techniques and provide breakthroughs with significant health benefits.

Some cloning of humans will be possible by 2015, but legal restrictions and public opinion may limit its actual extent. Controls are also likely for human modifications e.g., clone-based eugenic modifications for nondisease purposes. It is possible, however, that technology will enable genetic modifications for hereditary conditions i.e., sickle cell anemia through *in vitro* techniques or other mechanisms.

GMOs are also having a large effect on the scientific community as an enabling technology. Not only do "knock-out" animals with selected DNA sequences removed from their genome give scientists another tool to study the effect of the removed sequence on the animal, they also enable subsequent analysis of the interaction of those functions or components with the animal's entire system. Although knock-outs are not always complete, they provide another important tool to confirm or refute hypotheses regarding complex organisms.

Extant capabilities in genomics have already created opportunities yet have generated a number of issues. As more organisms are decoded and the functional implications of genes are discovered, concerns about property and privacy rights for the sequencing will likely continue.

The ability to profile an individual's DNA is already raising concerns about privacy and excessive monitoring. Examples include databases of DNA signatures for use in criminal investigations, and the potential use of genetically based health predispositions by insurance companies or employers to deny coverage or to discriminate. The latter may raise policy issues regarding acceptable and unacceptable profiling for insurance or employment. This issue is further worrisome because the exact code-to function mechanisms that trigger many disease predispositions are not well understood.

Issues may also arise if a strong genetic basis of human physical or cognitive ability is discovered. On the positive side, understanding a person's predisposition for certain abilities or limitations could enable custom educational or remediation programmes that will help to compensate for genetic inclinations, especially in early years when their effect can be optimized. On the negative side, groups may use such analyses in arguments to discriminate against target populations despite, the fact that ethnic distribution variances of cognitive ability are currently believed to be wider than ethnic mean differences, aggravating social and international conflicts.

Although the genetic profiles of plants have been modified for centuries using traditional techniques, questions regarding the safety of genetically modified foods have sparked international concerns in the United Kingdom and Europe, forcing a campaign by biotechnology companies to argue the safety of the technology and its applications. Some have argued that genetic engineering is actually as safe or safer than traditional combinatorial techniques such as irradiated seeds, since there often is strong supporting information concerning the function of the inserted sequences.

Governments have been forced into the issue, resulting in education efforts, food labeling proposals, and heated international trade discussions between the United States and Europe on the importation of GMOs and their seedlings. As genetic modification becomes more common, it may become more difficult to label and separate GMOs, resulting in a forcing function to resolve the issue of how far the technology should be applied and whether separate markets can be maintained in a global economy. This debate is starting to have global effects as populations in other countries begin to notice the impassioned debates in the United Kingdom and Europe.

Some have likened the anti-biotechnology movement to the anti-nuclear-power movement in scope and tactics, although the low cost and wide availability of basic genomic equipment and know-how will likely allow practically any country, small business, or even individual to participate in genetic engineering. Such wide technology availability and low entry costs could make it impossible for any movement or government to control the spread and use of genomic technology. At an extreme, successful protest pressures on big biotechnology companies together with wide technology availability could ultimately drive genomic engineering "underground" to groups outside such pressures and outside regulatory controls that help ensure safe and ethical uses. This could ironically facilitate the very problems that the anti-biotechnology movement is hoping to prevent.

Cloning and genetic modification also raise biodiversity concerns. Standardization of crops and livestock have already increased food supply vulnerabilities to diseases that can wipe out larger areas of production. Genetic modification may increase our ability to engineer responses to these threats, but the losses may still be felt in the production year unless broad-spectrum defences are developed.

In addition to food safety, the ability to modify biological organisms holds the possibility of engineered biological weapons that circumvent current or planned countermeasures. On the other hand, genomics could aid in biological warfare defence. Advances in genomics, therefore, could advance a race between threat engineering and countermeasures. Thus, although genetic manipulation is likely to result in medical advances, it is unclear whether we will be in a safer position in the future.

The rate at which GMO benefits are felt in poorer countries may depend on the costs of using patented organisms, marketing demands and approaches, and the rate at which crops become ubiquitous and inseparable from unmodified strains. Consider, current issues related to human immunodeficiency virus drug development and dissemination in poorer countries. Patentability has fueled research investments, but many poorer countries with dire needs cannot afford the latest drugs and must wait for handouts or patent expiration. Globalization, however, may fuel dissemination as multi-national companies invest in food production across the globe. Also, the rewards from opening previously unproductive land for production may provide the financial incentive to pay the premium for GMOs. Furthermore, widely available genomic technology could allow academics, nonprofit small businesses, and developing countries to develop GMOs to alleviate problems in poorer regions; larger biotechnology companies will focus on markets requiring capital-intensive R&D.

Finally, moral issues may play a large role in modulating the global effect of genomics trends. Some people simply believe it is improper to engineer or modify biological organisms using the new techniques. Unplanned side effects will likely support such opposition. Others are concerned with the real danger of eugenics programmes or of the engineering of dangerous biological organisms.

THERAPIES AND DRUG DEVELOPMENT

Technology

Beyond genetics, biotechnology will likely continue to improve therapies for preventing and treating disease and infection. New approaches might block a pathogen's ability to enter or travel in the body, leverage pathogen vulnerabilities, develop new countermeasure delivery mechanisms, or modulate or augment the immune response to recognizing new pathogens. These therapies may counter the current trend of increasing resistance to extant antibiotics, reshaping the war on infections.

In addition to addressing traditional viral and bacterial problems, therapies are being developed for chemical imbalances and modulation of chemical stasis. Antibodies are being developed that attack cocaine in the body and may be used to control addiction. Such approaches could have a significant effect on modifying the economics of the global illegal drug trade while improving conditions for users.

Drug development will likely be aided by various technology trends and enablers. Computer simulations combined with proliferating trends for molecular imaging technologies may continue to improve our ability to design molecules with desired functional properties that target specific receptors, binding sites, or markers, complementing combinatorial drug search with rational drug design. Simulations of drug interactions with target biological systems could become increasing useful in understanding drug efficacy and safety.

BIOMEDICAL ENGINEERING

Multidisciplinary teaming is accelerating advances and products in biomedical engineering and technology of organic and artificial tissues, organs, and materials.

Organic Tissues and Organs

Advances in tissue and organ engineering and repair are likely to result in organic and artificial replacement parts for humans. New advances in tissue regeneration and repair continue to improve our ability to resolve health problems within our bodies.

The field of tissue engineering, which is barely a decade old, has already led to engineered commercial skin products for wound treatment. Growth of cartilage for repair and replacement is at the stage of clinical testing, and treatment of heart disease via growth of functional tissue by 2015 is a realistic goal. These advances will depend upon improved biocompatible or bioabsorbable scaffold materials, development of 3D vascularized tissues and multicellular tissues, and an improved understanding of the *in vivo* growth process of cellular material on such scaffolds.

Research and applications of stem cell therapies will likely continue and expand, using these unspecialized human cells to augment or replace brain or body functions, organs, and structures. As the most unspecialized stem cells are found in early stage embryos or fetal tissue, an ethical debate is ensuing regarding the use of stem cells for research and therapy. Alternatives such as the use of adult human stem cells or stem cell culturing may ultimately produce large-scale cell supplies with reduced ethical concerns.

Xenotransplantations could be improved, aided by attempts to genetically modify donor tissue and organ antibodies, complements, and regulatory proteins to reduce or eliminate rejection. Baboons or pigs, may be genetically modified and cloned to produce organs for human transplant, although large-scale success may not occur by 2015.

Beyond rejection, the significance of xenotransplants is likely to be modulated by concerns that diseases such as retro viruses might jump from animals to people as a result of the transplantation techniques. Ethical and moral concerns as well as possible patenting issues may also result in regulations and limitations on xenotransplants, limiting their significance.

ARTIFICIAL MATERIALS, ORGANS, AND BIONICS

In addition to organic structures, advances are likely to continue in engineering artificial tissues and organs for humans.

Multi-functional materials are being developed that provide both structure and function or that have different properties on different sides, enabling new applications and capabilities. Polymers with a hydrophilic shell around a hydrophobic core can be used for timed release of hydrophobic drug molecules, as carriers for gene therapy or immobilized enzymes, or as artificial tissues. Sterically stabilized polymers could also be used for drug delivery.

Other materials are being developed for various biomedical applications. Fluorinated colloids, are being developed that take advantage of the high electronegativity of fluorine to enhance *in vivo* oxygen transport as a blood substitute during surgery and for drug delivery. Hydrogels with controlled swelling behaviour are being developed for drug delivery or as templates to attach growth materials for tissue engineering. Ceramics such as bioactive calcia-phosphate-silica glasses, hydroxyapetite, and calcium phosphates can serve as templates for bone growth and regeneration. Bioactive polymers can be applied as meshes, sponges, foams, or hydrogels to stimulate tissue growth. Coatings and surface treatments are being developed to increase biocompatibility of implanted materials to overcome the lack of endothelial cells in artificial blood vessels and reduce thrombosis. Blood substitutes may change the blood storage and retrieval systems while improving safety from blood-borne infections.

New manufacturing techniques and information technology are also enabling the production of biomedical structures with custom sizing and shape. It may become commonplace to manufacture custom ceramic replacement bones for injured hands, feet, and skull parts by combining computer tomography and "rapid prototyping" to reverse engineer new bones layer by layer.

Beyond structures and organs, neural and sensor prosthetics could begin to become significant by 2015. Retinas and cochlear implants, bypasses of spinal and other nerve damage, and other artificial communications and stimulations may improve and become more commonplace and affordable, eliminating many occurrences of blindness and deafness. This could eliminate or reduce the effect of serious handicaps and change society's response from accommodation to remediation.

BIOMIMETICS AND APPLIED BIOLOGY

Recent techniques such as functional brain imaging and knock-out animals are revolutionizing our endeavors to understand human and animal intelligence and capabilities. These efforts should, by 2015, make significant inroads in improving our understanding of phenomena such as false memories, attention, recognition, and information processing, with implications for better understanding people and designing and interfacing artificial systems such as autonomous robots and information systems. Neuromorphic engineering has already produced novel control algorithms, vision chips, head-eye systems, and biomimetic autonomous robots. Although not likely to produce systems with wide intelligence or capabilities similar to those of higher organisms, this trend may produce systems by 2015 that can robustly perform useful functions such as vacuuming a house, detecting mines, or conducting autonomous search.

Surgical and Diagnostic Biotechnology

Biotechnology and materials advances are likely to continue producing revolutionary surgical procedures and systems that will significantly reduce hospital stays and cost and increase effectiveness. New surgical tools and techniques and new materials and designs for vesicle and tissue support will likely continue to reduce surgical invasiveness and offer new solutions to medical problems. Techniques such as angioplasty may continue to eliminate whole classes of surgeries; others such as laser perforations of heart tissue could promote regeneration and healing. Advances in laser surgery could refine techniques and improve human capability, especially as costs are reduced and experience spreads. Hybrid imaging techniques will likely improve diagnosis, guide human and robotic surgery, and aid in basic understanding of body and brain function. Finally, collaborative information technology will likely extend specialized medical care to remote areas and aid in the global dissemination of medical quality and new advances.

By 2015, one can envision: effective localized, targeted, and controlled drug delivery systems; long-lived implants and prosthetics; and artificial skin, bone, and perhaps heart muscle or even nerve tissue. A host of social, political, and ethical issues such as those discussed above will likely accompany these developments.

Biomedical advances are already increasing human life span in countries where they are applied. New advances by 2015 are likely to continue this trend, accentuating issues such as shifts in population age demographics, financial support for retired persons, and increased health care costs for individuals. Advances, however, may improve not only life expectancy but productivity and utility of these individuals, offsetting or even overcoming the resulting issues.

Many costly and specialized medical techniques are likely to initially benefit citizens who can afford better medical care; wider global effects may occur later as a result of traditional trickle-down effects in medicine. Some technologies may have the opposite trend where low-cost technologies may enable cost-effective consulting with specialists regardless of location. However, access to technology may greatly mediate this dispersal mechanism and may place additional demands on technology upgrades and education. Countries that remain behind in terms of technological infrastructures may miss many of these benefits.

Theological debates have also raised concerns about the definition of what constitutes a human being, since animals are being modified to produce human organs for later xenotransplantation in humans. Genetic profiling may help to inform this debate as we understand the genetic differences between humans and animals.

Improved understanding of human intelligence and cognitive function could have broader legal and social effects. An understanding of false memories and how they are created could have an effect on legal liabilities and courtroom testimony. Understanding innate personal capabilities and job performance requirements could help us determine who would make better fighter pilots, who has an edge in analyzing complex images, and what types of improved training could improve people's capabilities to meet the special demands of their chosen careers. Ethical concerns could arise concerning discrimination against people who lack certain innate skills, requiring objective and careful measures for hiring and promotion.

Eventually, neural and sensory implants could radically change the way people sense, perceive, and interact with natural and artificial environments. Ultimately, these new capabilities could create new jobs and functions for people in these environments. Such innovations may first develop for individuals with particularly challenging and critical functions, but innovations may first develop in other quarters, given recent trends. Initial research indicates the feasibility of such implants and interactions, but it is unclear whether R&D and investments will accelerate enough to realise even such early applications by 2015. Current trends have concentrated on medical prosthetics where research prototypes are already appearing so it appears likely that globally significant systems will appear in this domain first.

THE PROCESS OF MATERIALS ENGINEERING

New materials can often be critical enabling drivers for new systems and applications with significant effects. However, it may not be obvious how enabling materials affect more observable trends and applications. A common process model from materials engineering can help to show how materials appear likely to break previous barriers in the process that ultimately results in applications with potential global benefits.

Developments in materials science and engineering result from interdisciplinary materials research. This development can be conveniently represented by the schematic description of the materials engineering process from concept to product/ application. This process view is a common approach in materials research circles and similar representations may be found in the literature.

Concept/Materials Design

Biomimetics is the design of systems, materials, and their functionality to mimic nature. Current examples include layering of materials to achieve the hardness of an abalone shell or trying to understand why spider silk is stronger than steel.

Combinatorial materials design uses computing power sometimes together with massive parallel experimentation to screen many different materials possibilities to optimize properties for specific applications catalysts, drugs, optical materials.

Composites are combinations of metals, ceramics, polymers, and biological materials that allow multi-functional behaviour. One common practice is reinforcing polymers or ceramics with ceramic fibres to increase strength while retaining light weight and avoiding the brittleness of the monolithic ceramic. Materials used in the body often combine biological and structural functions.

Nanoscale materials, i.e., materials with properties that can be controlled at submicrometer ($<10^{-6}$ m) or nanometer (10^{-9}

m) level, are an increasingly active area of research because properties in these size regimes are often fundamentally different from those of ordinary materials. Examples include carbon nanotubes, quantum dots, and biological molecules. These materials can be prepared either by purification methods or by tailored fabrication methods.

Processing, Properties, and Performance

These areas are inextricably linked to each other: Processing determines properties that in turn determine performance. Moreover, the sensitivity of instrumentation and measurement capability is often the enabling factor in optimizing processing, as for nanotechnology and micro electromechanical systems.

Rapid prototyping is the capability to combine computer-assisted design and manufacturing with rapid fabrication methods that allow inexpensive part production. Rapid prototyping enables a company to test several different inexpensive prototypes before committing infrastructure investments to an approach. Combined with manufacturing system improvements to allow flexibility of approach and machinery, rapid prototyping can lead to an *agile manufacturing* capability. Alternatively, the company can use its virtual capability to design and then outsource product manufacturing, thus offloading capital investment and risk. This capability is synergistic with the information technology revolution in the sense that it is a further factor in globalizing manufacturing capability and enabling organizations with less capital to have a significant technological effect. For the Department of Defence, it could reduce or eliminate requirements for warehousing large amounts of spares and, could enable the Air Force to "fly before they buy."

Self-assembly refers to the use in materials processing or fabrication of the tendency of some materials to organize themselves into ordered arrays. This provides a means to achieve structured materials "from the bottom up" as opposed to using manufacturing or fabrication methods such as lithography, which is limited by the measurement and instrumentation capabilities of the day. Organic polymers have been tagged with dye molecules to form arrays with lattice spacing in the visible optical wavelength range and that can be changed through chemical means. This provides a material that fluoresces and changes colour to indicate the presence of chemical species.

Manufacturing with DNA might represent the ultimate biomimetic manufacturing scheme. It consists of "functionalizing small inorganic building blocks with DNA and then using the molecular recognition processes associated with DNA to guide the assembly of those particles or building blocks into extended structures". Using this approach, Mirkin and colleagues demonstrated a highly selective and sensitive DNA-based chemical assay method using 13 nm diameter gold particles with attached DNA sequences. This approach is compatible with the commonly used polymerase chain reaction method of amplification of the amount of the target substance. *Micro- and nano-fabrication* methods include, lithography of coupled micro- or nano-scale devices on the same semiconductor or biological material. It is important to note the crucial role played in the development of these techniques by the parallel development of instrumentation and measurement devices such as the Atomic Force Microscope and the various Scanning Probe Microscopes.

Product/Application

The trends described above will likely work in concert to provide materials engineers with the capability to design and produce advanced materials that will be:

- *Smart*—Reactive materials combining sensors and actuators, perhaps together with computers, to enable response to environmental conditions and changes thereof. An example might be robots that mimic insects or birds for applications such as space exploration, hazardous materials location and treatment, and unmanned aerial vehicles.
- *Multi-functional*—MEMS and the "lab-on-a-chip" are excellent examples of systems that combine several

functions. Another example is a drug delivery system using a hydrogel with hydrophilic exterior and hydrophobic interior. Consider also aircraft skins fabricated from radar-absorbing materials that incorporate avionic links and the ability to modify shape in response to airflow.

- *Environmentally compatible or survivable*—The development of composite materials and the ability to tailor materials at the atomic level will likely provide opportunities to make materials more compatible with the environments in which they will be used. Examples might include prosthetic devices that serve as templates for the growth of natural tissue and structural materials that strengthen during service.

SMART MATERIALS

Several different types of materials exhibit sensing and actuation capabilities, including ferroelectrics, shape-memory alloys, and magnetostrictive materials. These effects also work in reverse, so that these materials, separately or together, can be used to combine sensing and actuation in response to environmental conditions. They are currently in widespread use in applications from ink-jet printers to magnetic disk drives to anti-coagulant devices.

An important class of smart materials is composites based upon lead zirconate titanate and related ferroelectric materials that allow increased sensitivity, multiple frequency response, and variable frequency. An example is the "Moonie"—a PZT transducer placed inside a half-moon-shaped cavity, which provides substantial amplification of the response. Another example is the use of composites of barium strontium titanate and non-ferroelectric materials that provide frequency-agile and field-agile responses. Applications include sensors and actuators that can change their frequency either to match a signal or to encode a signal. Ferroelectrics are already in use as nonvolatile memory elements for smart cards and as active elements in smart skis that change shape in response to stress.

Another important class of materials is smart polymers. Such electro-active polymers have already been used to make "artificial muscles". Currently available materials have limited mechanical power, but this is an active research area with potential applications to robots for space exploration, hazardous duty of various types, and surveillance. Hydrogels that swell and shrink in response to changes in pH or temperature are another possibility; these hydrogels could be used to deliver encapsulated drugs in response to changes in body chemistry. Another variation on this trend for controlled release of drugs is materials with hydrophilic exterior and hydrophobic interior.

A world with pervasive, networked sensors and actuators promises to improve, optimize, and customise the capability of systems and devices through availability of information and more direct actuation. Continuously available communication capability, ability to catalog and locate tagged personal items, and coordination of support functions have been espoused as benefits that may begin to be realised by 2015.

The continued development of small, low-profile biometric sensors, coupled with research on voice, handwriting, and fingerprint recognition, could provide effective personal security systems. These could be used for identification by police/military and also in business, personal, and leisure applications. Combined with today's information technologies, such uses could help resolve nagging security and privacy concerns while enabling other applications such as improved handgun safety and vehicle theft control.

Other potential applications of smart materials that would be enabled by 2015 include: clothes that respond to weather, interface with information systems, monitor vital signs, deliver medicines, and automatically protect wounds; airfoils that respond to airflow; buildings that adjust to the weather; bridges and roads that sense and repair cracks; kitchens that cook with wireless instructions; virtual reality telephones and entertainment centres; and personal medical diagnostics. The level of development and integration of these technologies into everyday life will probably depend more on consumer attitudes than on technical developments.

In addition to the surveillance and identification functions mentioned under smart materials above, developments in robotics may provide new and more sensitive capabilities for detecting and destroying explosives and contraband materials and for operating in hazardous environments. Increases in materials performance, both for power sources and for sensing and actuation, as well as integration of these functions with computing power, could enable these applications.

Such trend potentials are not without issues. Pervasive sensory information and access to collected data raise significant privacy concerns. Also, the pace of development will likely depend on investment levels and market drivers. In many cases the immediate benefits and cost savings from smart material applications will continue to drive development, but more exotic materials research may depend on public commitment to research and belief in investing in longer-term rewards.

Examples of self-assembling materials include colloidal crystal arrays with mesoscale lattice constants that form optical diffraction gratings, and thus change colour as the array swells in response to heat or chemical changes. In the case of a hydrogel with an attached side group that has molecular recognition capability, this is a chemical sensor. Self-assembling colloidal suspensions have been used to form a light-emitting diode, a porous metal array, and a molecular computer switch.

The DNA-based self-assembly mentioned above was achieved by attaching non-linking DNA strands to metal nanoparticles and adding a linking agent to form a DNA lattice. This can be turned into a biosensor or a nanolithography technique for biomolecules.

NANOMATERIALS

This area combines nanotechnology and many applications of nanostructured materials. One important research area is the formation of semiconductor "quantum dots" by injecting precursor materials conventionally used for chemical-vapour deposition of semiconductors into a hot liquid surfactant. This "quantum dot" is in reality a macromolecule because it is coated with a monolayer of the surfactant, preventing agglomeration. These materials photoluminesce at different frequencies depending upon their size, allowing optical multiplexing in biological labeling.

Another important class of nanomaterials is nanotubes. Possible applications are field-emission displays, nanoscale wires for batteries, storage of Li or H_2, and thermal management. Another possibility is to use nanotubes as reinforcement for composite materials. Presumably because of the nature of the bonding, it is predicted that nanotube-based material could be 50 to 100 times stronger than steel at one-sixth of the weight if current technical barriers can be overcome.

Nanoscale structures with desirable mechanical and other properties may also be obtained through processing. Examples include strengthening of alloys with nanoscale grain structure, increased ductility of metals with multi-phase nanoscale microstructure, and increased flame retardancy of plastic nanocomposites.

NANOTECHNOLOGY

Much has been made of the trend toward producing devices with ever-decreasing scale. Many people have projected that nanometer-scale devices will continue this trend, bringing it to unprecedented levels. This includes scale reduction not only in microelectronics but also in fields such as MEMS and quantum-switch-based computing in the shorter term. These advances have the potential to change the way we engineer our environment, construct and control systems, and interact in society.

Nanofabricated Chips

SEMATECH—the leading industry group in the semiconductor manufacturing business—is calling for the development of nanoscale semiconductors in their latest International Technology Roadmap for Semiconductors. The roadmap calls for a 35 nm gate length in 2015 with a total number of functions in high-volume production microprocessors of around 4.3 billion. For low-volume, high-performance processors, the number of functions may approach 20 billion. Corresponding memory chips are targeted to hold around 64 gigabytes. These roadmap targets would continue the exponential trend in processing power, fueling advances in information technology. Although a number of engineering challenges exist, obstacles to achieve at least this level of performance do not seem insurmountable.

Given unforeseen shortfalls in the economic production of these chips, several alternatives seem possible. Defect-tolerant computer architectures such as those prototyped on a small scale by Hewlett-Packard offer one alternative. These alternative methods provide some level of additional robustness to the performance goals set by the ITRS.

However, in the years following 2015, additional difficulties will likely be encountered, some of which may pose serious challenges to traditional semiconductor manufacturing techniques. In particular, limits to the degree that interconnections or "wires" between transistors may be scaled could in turn limit the effective computation speed of devices because of materials properties and compatibility, despite incremental present-day advances in these areas. Thermal dissipation in chips with extremely high device densities will also pose a serious challenge. This issue is not so much a fundamental limitation as it is an economic consideration, in that heat dissipation mechanisms and cooling technology may be required that add to total system cost, thereby adversely affecting marginal cost per computational function for these devices.

Quantum-Switch-Based Computing

One potential long-term solution for overcoming obstacles to increased computational power is computing based on devices that take advantage of various quantum effects. The core innovation in this work is the use of quantum effects, such as spin polarization of electrons, to determine the state of individual switches. This is in contrast to more traditional microelectronics, which are based on macroscopic properties of large numbers of electrons, taking advantage of materials properties of semiconductors.

Various concepts of quantum computers are attractive because of their massive parallelism in computation, but they are not anticipated to have significant effects by 2015. These concepts are qualitatively different from those employed in traditional computers and will hence require new computer architectures. The types of computations that can be quickly performed using these computers are not the same as those readily addressed by today's digital computer. Several workers in the area have devised algorithms for problems that are very computationally intensive for existing digital computers, which could be made much faster using the physics of quantum computers. Examples of these problems include factoring large numbers, searching large databases, pattern matching, and simulation of molecular and quantum phenomena.

A preliminary survey of work in this area indicates that quantum switches are unlikely to overcome major technical obstacles, such as error correction, de-coherence and signal input/output, within the next 15 years. If this were indeed the case, quantum-switch-based computing does not appear to be competitive with traditional digital electronic computers within the 2015 timeframe.

BIO-MOLECULAR DEVICES AND MOLECULAR ELECTRONICS

Many of the same manufacturing and architectural challenges discussed above regarding quantum computing also hold true for molecular electronics. Molecular electronic devices could operate as logic switches through chemical means, using synthesized organic compounds. These devices can be assembled chemically in large numbers and organized to form a computer. The main advantage of this approach is significantly lower power consumption by individual devices. Several approaches for such devices have been devised, and experiments have shown evidence of switching behaviour for individual devices. Several research groups have proposed interconnection between devices using carbon nanotubes, which provide high conductivity using single molecular strands of carbon. Progress has been made toward raising the operating temperature of these switches to nearly room temperature, making the switching process reversible, and increasing the overall amount of current that can be switched using these devices.

Several major outstanding issues remain with respect to molecular electronics. One issue is that molecular memories must be able to maintain their state, just as in a digital electronic computer. Also, given that the manufacturing and assembly process for these devices will lead to device defects, a defect-tolerant computer architecture needs to be developed. Fabricating reliable interconnects between devices using carbon nanotubes is an additional challenge. A significant amount of work is ongoing in each of these areas. Even though experimental progress to date in this area has been substantial, it seems unlikely that molecular computers could be developed within the next 15 years that would be relatively attractive compared with conventional electronic computers.

Examining the potential for developing qualitatively different computational capabilities from different technology bases is a challenging exercise. The history of computing over the last 50 years has seen one major shift in technology base, with a corresponding shift not just in computational power but also in attitudes about the value of computers. Ideas of computers as simple machines for computation gave way to the use of computers for personal productivity with the advent of the microprocessor. As the power of these microprocessors has grown exponentially, they have also been seen more recently as a vehicle for new media and socialization.

The ramifications of future computing technologies will be determined principally by two factors: the conception, development, and adoption of new applications that require significantly more computational power; and the ability of technology to address these demands. New applications are always difficult to anticipate, but it is less challenging to foresee the likely consequences for diffusion of this technology. Past experience with personal computers and telecommunications has shown that these technologies diffuse more rapidly in the developed world than in the developing world. It is difficult to foresee an increase in the political or ethical barriers to computing technologies beyond those seen today, and these are rapidly vanishing.

On the remaining question of technology development, the odds-on favourite for the next 15 years remains traditional digital electronic computers based on semiconductor technology. Given the virtual certainty of continued progress in this area, it is hard to imagine a scenario in which a competing technology could offer a significant performance advantage at a competitive price. But the longer-term, traditionally elusive question in the period after 2015 is: How long will traditional silicon computing last? And when, if ever, will a competing technology become available and attractive? If an alternative computing technology becomes sufficiently attractive, the economic effects of technology substitution on the current semiconductor industry and adjacent industries must be considered. Major industry players may be faced with a choice between cannibalizing their existing market opportunities in favour of these new, future technologies, competing head-on with new players, or simply acquiring them. Most important, given the very different architectural approaches of these technologies and the classes of problems for which they are best suited, what will be the effect on future applications? The promises of nanotechnology may indeed become a reality in the period after 2015, but it will face these competitive challenges before its significance becomes global.

INTEGRATED MICROSYSTEMS AND MEMS

MEMS is less an application area in itself than a manufacturing or fabrication technique that enables other application areas. Many authors use MEMS as shorthand to imply a number of particular application areas. As it is used here, MEMS is a "topdown" fabrication technology that is especially useful for integrating mechanical and electrical systems together on the same chip. It is grouped in the category of integrated microsystems because these same MEMS techniques can be extended in the future to also help integrate biological and chemical components on the same chip. Thus far, MEMS techniques have been used to make some functional commercial devices such as sensors and single-chip measurement devices. Many researchers have used MEMS technologies as analytical tools in other areas of nanotechnology such as the ones discussed here.

Smart Systems-on-a-Chip

Simple electro-optical and chemical sensor components have already been successfully integrated onto logic and memory chip designs in research and development labs. Likewise, radio frequency component integration in wireless devices is already being produced in mass quantities. Some companies have products capable of doing elementary DNA testing. The 1999 ITRS predicts the introduction of chemical sensor components with logic in commercial designs by 2002, with electro-optical component integration by 2004, and biological systems integration by 2006. Given these predictions, there is clearly time for relatively complex integrated systems and applications to develop within the 2015 timeline. These advances could enable many applications where increased integrated functionality can become ubiquitous as a result of lower costs and micro-packaging.

Nanoscale Instrumentation and Measurement Technology

Instrumentation and measurement technologies are some of the most promising areas for near-term advancements and enabling effects. As optical, fluidic, chemical, and biological components can be integrated with electronic logic and memory components on the same chip at marginal cost, drug discovery, genetics research, chemical assays, and chemical synthesis are all likely to be substantially affected by these advances by 2015.

Some of the first applications of nanoscale instruments were as basic sensors for acceleration, pressure, etc. Small, microscale, special-purpose optical and chemical sensors have been used for some time in sophisticated laboratory equipment, along with microprocessors for signal processing and computation. Already, companies have produced products that allow for basic DNA analysis, and that can assist in drug discovery. As these sensors become more sophisticated and more integrated with computational capability, their utility should grow tremendously, especially in the biomedical arena.

There are several advantages of nanotechnology for integrated systems in general and instrumentation and measurement systems as a subset of these. First, existing semiconductor technologies will likely allow the volume manufacture of integrated smart systems that can be produced at low enough cost to be considered disposable. Second, the massive parallelism afforded by this same technology allows for the rapid analysis with integrated computation of very complex samples such as DNA, the processing of large numbers of samples, and the recognition of large numbers of agents. Devices with these properties are already of tremendous utility in the biomedical arena for drug testing, chemical assays, etc. In addition, they will likely find utility in a variety of industrial applications.

Nanosystems are already starting to affect applications where miniaturization of components, subsystems, and even complete systems is significantly reducing device size, power, and consumables while introducing new capabilities. This area lends itself naturally to the confluence of all the broad areas discussed in this report. The next five to ten years will likely see the integration of computational capabilities with biological, chemical, and optical components in systems-on-a-chip. At the same time, advances in biotechnology should drive applications for drug discovery and genomics, as well as the basic understanding of many other phenomena. Advances in biomaterials will likely produce biologically compatible packaging, capable of isolating substances from the body in a time-controlled fashion. The confluence of these capabilities could allow for continued development of microscale and nanoscale systems that could continue to be introduced into the body to perform basic diagnostic functions in a minimally invasive way, providing new abilities to remedy health problems.

Other possible applications include: pervasive, self-moving sensor systems; nanoscrubbers and nanocatalysts; even inexpensive, networked "nanosatellites." So-called "nanosatellites" are targeting order-of-magnitude reductions in both size and mass by reducing major system components using integrated microsystems. If successful, this could economize current missions and approaches while enabling new missions. In addition, advances could empower the proliferation of currently controlled processing capabilities with associated threats to international security. Progress will likely depend on investment levels as well as continued S&T development and progress.

MOLECULAR MANUFACTURING AND NANOROBOTS

A number of experts have put forth the concept of molecular manufacturing where objects are assembled atom by atom or molecule by molecule. Bottom-up molecular manufacturing differs from microtechnology and MEMS in that the latter employ top-down approaches using bulk materials using macroscopic fabrication techniques.

To realise molecular manufacturing, a number of technical accomplishments are necessary. First, suitable molecular building blocks must be found. These building blocks must be physically durable, chemically stable, easily manipulated, and to a certain extent functionally versatile. Several workers in the field have suggested the use of carbon-based diamond-like structures as building blocks for nano-mechanical devices, such as gears, pivots, and rotors. Other molecules could also be used to build structures, and to provide other integrated capabilities, such as chemically reactive structures. Much additional work in the area of modeling and synthesis of appropriate molecular structures is needed, and a number of groups are working to this end. Dresselhaus and others are fabricating suitable molecular building blocks for these structures.

The second major area for development is in the ability to assemble complex structures based on a particular design. A number of researchers have been working on different approaches to this issue. Different techniques for physical placements are under development. One approach by Quate, MacDonald, and Eigler uses atomicforce or molecular microscopes with very small nanoprobes to move atoms or molecules around with the aid of physical or chemical forces. An alternative approach by Prentiss uses lasers to place molecules in a desired location. Chemical assembly techniques are being addressed by a number of groups, including Whiteside's approach to building structures one molecular layer at a time.

A third major area for development within molecular manufacturing is systems design and engineering. Extremely complex molecular systems at the macro scale will require substantial subsystem design, overall system design, and systems integration, much like complex manufactured systems of the present day. Although the design issues are likely to be largely separable at a subsystems level, the amount of computation required for design and validation is likely to be quite substantial. Performing checks on engineering constraints, such as defect tolerance, physical integrity, and chemical stability, will be required as well.

Some workers in the area have outlined a potential path for the evolution of molecular manufacturing capability, which is broken down by overall size, type of fabrication technology, system complexity, component materials used, etc. Some versions of this concept foresee the use of massively parallel nanorobots or scanning nanoprobes to assemble structures physically. Other, more advanced concepts incorporate chemical principles and use simple chemical feedstocks to achieve much larger devices on the order of 10^8 to 10^9 molecular parts.

Ostensibly, as each of these techniques matures, more systems and engineering-level work must be done before applications can be realised on a significant scale. Although molecular manufacturing holds the promise of significant global changes such as retraining large numbers of manufacturing workforces, opportunities for new regions to vie for dominance in a new manufacturing paradigm, or a shift to countries that do not have legacy manufacturing infrastructures, it remains the least concrete of the technologies discussed here. Significant progress has been made, however, in the development of component technologies within the first regime of molecular manufacturing, where objects might be constructed from simple molecules and manufactured in a short amount of time via parallel atomic force microprobes or from simple self-assembled structures. Although the building blocks for these systems currently exist only in isolation at the research stage, it is certainly reasonable to expect that an integrated capability could be developed over the next 15 years. Such a system could be able to assemble structures with between 100 and 10,000 components and total dimensions of perhaps tens of microns. A series of important breakthroughs could certainly cause progress in this area to develop much more rapidly, but it seems very unlikely that macro-scale objects could be constructed using molecular manufacturing within the 2015 timeframe.

6

Nanosensors and Nanoscale Scanning

NANOSENSOR

Nanosensors are any biological, chemical, or physical sensory points used to convey information about nanoparticles to the macroscopic world. Though humans have not yet been able to synthesize nanosensors, predictions for their use mainly include various medicinal purposes and as gateways to building other nanoproducts, such as computer chips that work at the nanoscale and nanorobots. Presently, there are several ways proposed to make nanosensors, including top-down lithography, bottom-up assembly, and molecular self-assembly.

Predicted Applications

Medicinal uses of nanosensors mainly revolve around the potential of nanosensors to accurately identify particular cells or places in the body in need. By measuring changes in volume, concentration, displacement and velocity, gravitational, electrical, and magnetic forces, pressure, or temperature of cells in a body, nanosensors may be able to distinguish between and recognize certain cells, most notably those of cancer, at the molecular level in order to deliver medicine or monitor development to specific places in the body. In addition, they may be able to detect macroscopic variations from outside the body and communicate these changes to other nanoproducts working within the body.

One example of nanosensors involves using the fluorescence properties of cadmium selenide quantum dots as sensors to uncover tumors within the body. By injecting a body with these quantum dots, a doctor could see where a tumor or cancer cell was by finding the injected quantum dots, an easy process because of their fluorescence. Developed nanosensor quantum dots would be specifically constructed to find only the particular cell for which the body was at risk. A downside to the cadmium selenide dots, however, is that they are highly toxic to the body. As a result, researchers are working on developing alternate dots made out of a different, less toxic material while still retaining some of the fluorescence properties. In particular, they have been investigating the particular benefits of zinc sulfide quantum dots which, though they are not quite as fluorescent as cadmium selenide, can be augmented with other metals including manganese and various lanthanide elements. In addition, these newer quantum dots become more fluorescent when they bond to their target cells. Potential predicted functions may also include sensors used to detect specific DNA in order to recognize explicit genetic defects, especially for individuals at high-risk and implanted sensors that can automatically detect glucose levels for diabetic subjects more simply than current detectors. DNA can also serve as sacrificial layer for manufacturing CMOS IC, integrating a nanodevice with sensing capabilities. Therefore, using proteomic patterns and new hybrid materials, nanobiosensors can also be used to enable components configured into a hybrid semiconductor substrate as part of the circuit assembly. The development and miniaturization of nanobiosensors should provide interesting new opportunities.

Other projected products most commonly involve using nanosensors to build smaller integrated circuits, as well as incorporating them into various other commodities made using other forms of nanotechnology for use in a variety of situations including transportation, communication, improvements in structural integrity, and robotics. Nanosensors may also eventually be valuable as more accurate monitors of material states for use in systems where size and weight are constrained, such as in satellites and other aeronautic machines.

Existing Nanosensors

Currently, the most common mass-produced functioning nanosensors exist in the biological world as natural receptors of outside stimulation. Sense of smell, especially in animals in which it is particularly strong, such as dogs, functions using receptors that sense nanosized molecules. Certain plants, too, use nanosensors to detect sunlight; various fish use nanosensors to detect minuscule vibrations in the surrounding water; and many insects detect sex pheromones using nanosensors.

Certain electromagnetic sensors have also been in use in photoelectric systems. These work because the specific sensors called, aptly, photosensors are easily influenced by light of various wavelengths. The electromagnetic source transfers energy to the photosensors and energizes them into an excited state which causes them to release an electron into a semiconductor. At that point, it is relatively easy to detect the electricity coming from the sensors, and thus easy to know if the sensors are receiving light. Though more advanced uses of photosensors incorporating other forms of nanotechnology have yet to be implemented into consumer society, most film cameras have used photosensors at the nano size for years. Traditional film uses a layer of silver ions that become excited by solar energy and clump into groups, as small as four atoms apiece in some cases, that scatter light and appear dark on the frame. Various other types of film can be made using a similar process to detect other specific wavelengths of light, including x-rays, infrared, and ultraviolet.

One of the first working examples of a synthetic nanosensor was built by researchers at the Georgia Institute of Technology in 1999. It involved attaching a single particle onto the end of a carbon nanotube and measuring the vibrational frequency of the nanotube both with and without the particle. The discrepancy between the two frequencies allowed the researchers to measure the mass of the attached particle.

Chemical sensors, too, have been built using nanotubes to detect various properties of gaseous molecules. Carbon nanotubes have been used to sense ionization of gaseous molecules while nanotubes made out of titanium have been employed to detect atmospheric concentrations of hydrogen at the molecular level. Many of these involve a system by which nanosensors are built to have a specific pocket for another molecule. When that particular molecule, and only that specific molecule, fits into the nanosensor, and light is shone upon the nanosensor, it will reflect different wavelengths of light and, thus, be a different colour.

Production Methods

There are currently several hypothesized ways to produce nanosensors. Top-down lithography is the manner in which most integrated circuits are now made. It involves starting out with a larger block of some material and carving out the desired form. These carved out devices, notably put to use in specific microelectromechanical systems used as microsensors, generally only reach the micro size, but the most recent of these have begun to incorporate nanosized components.

Another way to produce nanosensors is through the bottom-up method, which involves assembling the sensors out of even more minuscule components, most likely individual atoms or molecules. This would involve moving atoms of a particular substance one by one into particular positions which, though it has been achieved in laboratory tests using tools such as atomic force microscopes, is still a significant difficulty, especially to do en masse, both for logistic reasons as well as economic ones. Most likely, this process would be used mainly for building starter molecules for self-assembling sensors.

he third way, which promises far faster results, involves self-assembly, or "growing" particular nanostructures to be used as sensors. This most often entails one of two types of assembly. The first involves using a piece of some previously created or naturally formed nanostructure and immersing it in free atoms of its own kind. After a given period, the structure, having an irregular surface that would make it prone to attracting more molecules as a continuation of its current pattern, would capture some of the free atoms and continue to form more of itself to make larger components of nanosensors.

The second type of self-assembly starts with an already complete set of components that would automatically assemble themselves into a finished product. Though this has been so far successful only in assembling computer chips at the micro size, researchers hope to eventually be able to do it at the nanometer size for multiple products, including nanosensors. Accurately being able to reproduce this effect for a desired sensor in a laboratory would imply that scientists could manufacture nanosensors much more quickly and potentially far more cheaply by letting numerous molecules assemble themselves with little or no outside influence, rather than having to manually assemble each sensor.

WHAT ARE NANOROBOTS

Nanorobotics is emerging as a demanding field dealing with miniscule things at molecular level. Nanorobots are quintessential nanoelectromechanical systems designed to perform a specific task with precision at nanoscale dimensions. Its advantage over conventional medicine lies on its size. Particle size has effect on serum lifetime and pattern of deposition. This allows drugs of nanosize to be used in lower concentration and has an earlier onset of therapeutic action. It also provides materials for controlled drug delivery by directing carriers to a specific location. The typical medical nanodevice will probably be a micron-scale robot assembled from nanoscale parts. These nanorobots can work together in response to environment stimuli and programmed principles to produce macro scale results.

Elements of Nanorobots

Carbon will likely be the principal element comprising the bulk of a medical nanorobot, probably in the form of diamond or diamondoid/fullerene nanocomposites. Many other light elements such as hydrogen, sulfur, oxygen, nitrogen, fluorine, silicon, etc. will be used for special purposes in nanoscale gears and other components. The chemical inertness of diamond is proved by several experimental studies. One such experiment conducted on mouse peritoneal macrophages cultured on DLC showed no significant excess release of lactate dehydrogenase or of the lysosomal enzyme beta N-acetyl-D-glucosaminidase an enzyme known to be released from macrophages during inflammation.

Morphological examination revealed no physical damage to either fibroblasts or macrophages, and human osteoblast like cells confirming the biochemical indication that there was no toxicity and that no inflammatory reaction was elicited in vitro. The smoother and more flawless the diamond surface, the lesser is the leukocyte activity and fibrinogen adsorption. Interestingly, on the rougher "polished" surface, a small number of spread and fused macrophages were present, indicating that some activation had occurred. The exterior surface with near-nanometer smoothness results in very low bioactivity. Due to the extremely high surface energy of the passivated diamond surface and the strong hydrophobicity of the diamond surface, the diamond exterior is almost completely chemically inert.

Constituents and Design of Nanorobots

Nanorobots will possess full panoply of autonomous subsystems whose design is derived from biological models. Drexler evidently was the first to point out, in 1981, that complex devices resemble biological models in their structural components. The various components in the nanorobot design may include onboard sensors, motors, manipulators, power supplies, and molecular computers. Perhaps the best-known biological example of such molecular machinery is the ribosome the only freely programmable nanoscale assembler already in existence. The mechanism by which protein binds to the specific receptor site might be copied to construct the molecular robotic arm.

The manipulator arm can also be driven by a detailed sequence of control signals, just as the ribosome needs mRNA to guide its actions. These control signals are provided by external acoustic, electrical, or chemical signals that are received by the robot arm via an onboard sensor using a simple "broadcast architecture a technique which can also be used to import power. the biological cell may be regarded as an example of a broadcast architecture in which the nucleus of the cell send signals in the form of mRNA to ribosomes in order to manufacture cellular proteins.

Assemblers are molecular machine systems that could be described as systems capable of performing molecular manufacturing at the atomic scale which require control signals provided by an onboard nanocomputer This programmable nanocomputer must be able to accept stored instructions which are sequentially executed to direct the manipulator arm to place the correct moiety or nanopart in the desired position and orientation, thus giving precise control over the timing and locations of chemical reactions or assembly operations.

Construction of Nanorobots

There are two main approaches to building at the nanometer scale: positional assembly and self-assembly. In positional assembly, investigators employ some devices such as the arm of a miniature robot or a microscopic set to pick up molecules one by one and assemble them manually. In contrast, self-assembly is much less painstaking, because it takes advantage of the natural tendency of certain molecules to seek one another out. With self-assembling components, all that investigators have to do is put billions of them into a beaker and let their natural affinities join them automatically into the desired configurations. Making complex nanorobotic systems requires manufacturing techniques that can build a molecular structure via computational models of diamond mechanosynthesis. DMS is the controlled addition of carbon atoms to the growth surface of a diamond crystal lattice in a vacuum-manufacturing environment. Covalent chemical bonds are formed one by one as the result of positionally constrained mechanical forces applied at the tip of a scanning probe microscope apparatus, following a programmed sequence.

Recognition of Target Site by Nanorobots

Different molecule types are distinguished by a series of chemotactic sensors whose binding sites have a different affinity for each kind of molecule. The control system must ensure a suitable performance. It can be demonstrated with a determined number of nanorobots responding as fast as possible for a specific task based scenario. In the 3D workspace the target has surface chemicals allowing the nanorobots to detect and recognize it. Manufacturing better sensors and actuators with nanoscale sizes makes them find the source of release of the chemical. Nanorobot Control Design simulator was developed, which is software for nanorobots in environments with fluids dominated by Brownian motion and viscous rather than inertial forces.

First, as a point of comparison, the scientists used the nanorobots' small Brownian motions to find the target by random search. In a second method, the nanorobots monitor for chemical concentration significantly above the background level. After detecting the signal, a nanorobot estimates the concentration gradient and moves toward higher concentrations until it reaches the target. In the third approach, nanorobots at the target release another chemical, which others use as an additional guiding signal to the target. With these signal concentrations, only nanorobots passing within a few microns of the target are likely to detect the signal.

Thus, we can improve the response by having the nanorobots maintain positions near the vessel wall instead of floating throughout the volume flow in the vessel from monitoring the concentration of a signal from others; a nanorobot can estimate the number of nanorobots at the target. So, the nanorobot uses this information to determine when enough nanorobots are at the target, thereby terminating any additional "attractant" signal a nanorobot may be releasing. It is found that the nanorobots stop attracting others once enough nanorobots have responded. The amount is considered enough when the target region is densely covered by nanorobots. Thus these tiny machines work at the target site accurately and precisely to that extent only to which it is designed to do.

Nanorobots Evading the Immune System

Every medical nanorobot placed inside the human body will encounter phagocytic cells many times during its mission. Thus all Nanorobots, which are of a size capable of ingestion by phagocytic cells, must incorporate physical mechanisms and operational protocols for avoiding and escaping from phagocytes. The initial strategy for medical nanorobots is first to avoid phagocytic contact or recognition. To avoid being attacked by the host's immune system, the best choice is to have an exterior coating of passive diamond. The smoother and flawless the coating, the lesser is the reaction from the body's immune system. And if this fails then to avoid it's binding to the phagocyte surface that leads to phagocytic activation. If trapped, the medical nanorobot can induce exocytosis of the phagosomal vacuole in which it is lodged or inhibit both phagolysosomal fusion and phagosome metabolism.

In rare circumstances, it may be necessary to kill the phagocyte or to blockade the entire phagocytic system. The most direct approach for a fully functional medical nanorobot is to employ its motility mechanisms to locomote out of, or away from, the phagocytic cell that is attempting to engulf it. This may involve reverse cytopenetration, which must be done cautiously. It is possible that frustrated phagocytosis may induce a localized compensatory granulomatous reaction. Medical nanorobots therefore may also need to employ simple but active defensive strategies to forestall granuloma formation. Metabolizing local glucose and oxygen for energy can do the powering of the nanorobots. In a clinical environment, another option would be externally supplied acoustic energy. When the task of the nanorobots is completed, they can be retrieved by allowing them to exfuse themselves via the usual human excretory channels or can also be removed by active scavenger systems.

Nanorobots in Cancer Detection and Treatment

The development of nanorobots may provide remarkable advances for diagnosis and treatment of cancer. Nanorobots could be a very helpful and hopeful for the therapy of patients, since current treatments like radiation therapy and chemotherapy often end up destroying more healthy cells than cancerous ones. From this point of view, it provides a non-depressed therapy for cancer patients. The Nanorobots will be able to distinguish between different cell types that is the malignant and the normal cells by checking their surface antigens. This is accomplished by the use of chemotactic sensors keyed to the specific antigens on the target cells. Another approach uses the innovative methodology to achieve decentralized control for a distributed collective action in the combat of cancer. Using chemical sensors they can be programmed to detect different levels of E-cadherin and beta-catenin in primary and metastatic phases. Medical nanorobots will then destroy these cells, and only these cells. The following control methods were considered:

- *Random:* Nanorobots moving passively with the fluid reaching the target only if they bump into it due to rownian motion.
- *Follow gradient:* Nanorobots monitor concentration intensity for E-cadherin signals, when detected, measure and follow the gradient until reaching the target. If the gradient estimate subsequent to signal detection finds no additional signal in50ms, the nanorobot considers the signal to be a false positive and continues flowing with the fluid.

- Follow gradient with attractant: as above, but nanorobots arriving at the target, they release in addition a different chemical signal used by others to improve their ability to find the target. Thus, a higher gradient of signal intensity of E-cadherin is used as chemical parameter identification in guiding nanorobots to identify malignant tissues. Integrated nanosensors can be utilized for such a task in order to find intensity of E-cadherin signals. Thus they can be employed effectively for treating cancer.

Nanorobots IN Cancer Detection and Treatment

Pharmacyte is a self-powered, computer controlled medical nanorobot system capable of digitally precise transport, timing, and targeted-delivery of pharmaceutical agents to specific cellular and intracellular destinations within the human body. Pharmacytes escape the phagocytic process as they will not embolize small blood vessels because the minimum viable human capillary that allows passage of intact erythrocytes and white cells is 3–4 micronmeter in diameter, which is larger than the largest proposed Pharmacyte.

Pharmacytes will have many applications in nanomedicine such as initiation of apoptosis in cancer cells and direct control of cell signaling processes. Pharmacytes could also tag target cells with biochemical natural defensive or scavenging systems, a strategy called "phagocytic flagging". Novel recognition molecules are expressed on the surface of apoptotic cells. In the case of T lymphocytes, one such molecule is phosphatidylserine, a lipid that is normally restricted to the inner side of the plasma membrane [1m] but, after the induction of apoptosis, appears on the outside.

Cells bearing this molecule on their surface can then be recognized and removed by phagocytic cells. Seeding the outer wall of a target cell with phosphatidylserine or other molecules with similar action could activate phagocytic behaviour by macrophages, which had mistakenly identified the target cell as apoptotic substances capable of triggering a reaction by the body Pharmacytes would be capable of carrying up to approximately 1cubicmeter of pharmaceutical payload stored in onboard tanks that are mechanically offloaded using molecular sorting pumps operated under the control of an onboard computer.

Depending on mission requirements, the payload can be discharged into the proximate extracellular fluid or delivered directly into the cytosol using a transmembrane injector mechanism. If needed for a particular application, deployable mechanical cilia and other locomotive systems can be added to the Pharmacyte to permit transvascular and transcellular mobility, thus allowing delivery of pharmaceutical molecules to specific cellular and even intracellular addresses with negligible error. Pharmacytes, once depleted of their payloads or having completed their mission, would be recovered from the patient by conventional excretory pathways. The nanorobots might then be recharged, reprogrammed and recycled for use in a second patient who may need a different pharmaceutical agent targeted to different tissues or cells than in the first patient.

Nanorobots in the Diagnosis and Treatment of Diabetes

Glucose carried through the blood stream is important to maintain the human metabolism working healthfully, and its correct level is a key issue in the diagnosis and treatment of diabetes. Intrinsically related to the glucose molecules, the protein hSGLT3 has an important influence in maintaining proper gastrointestinal cholinergic nerve and skeletal muscle function activities, regulating extracellular glucose concentration. The hSGLT3 molecule can serve to define the glucose levels for diabetes patients. The most interesting aspect of this protein is the fact that it serves as a sensor to identify glucose.

The simulated nanorobot prototype model has embedded Complementary Metal Oxide semi-conductor nanobioelectronics. It features a size of ~2 micronmeter, which permits it to operate freely inside the body. Whether the nanorobot is invisible or visible for the immune reactions, it has no interference for detecting glucose levels in blood stream.

Even with the immune system reaction inside the body, the nanorobot is not attacked by the white blood cells due biocompatibility. For the glucose monitoring the nanorobot uses embedded chemosensor that involves the modulation of hSGLT3 protein glucosensor activity.

Through its onboard chemical sensor, the nanorobot can thus effectively determine if the patient needs to inject insulin or take any further action, such as any medication clinically prescribed. The image of the NCD simulator workspace shows the inside view of a venule blood vessel with grid texture, red blood cells and nanorobots. They flow with the RBCs through the bloodstream detecting the glucose levels. At a typical glucose concentration, the nanorobots try to keep the glucose levels ranging around 130 mg/dl as a target for the Blood Glucose Levels. A variation of 30mg/dl was adopted as a displacement range, though this can be changed based on medical prescriptions. In the medical nanorobot architecture, the significant measured data can be then transferred automatically through the RF signals to the mobile phone carried by the patient. At any time, if the glucose achieves critical levels, the nanorobot emits an alarm through the mobile phone.

Controlling Glucose Level using Nanorobots

In the simulation, the nanorobot is programmed also to emit a signal based on specified lunch times, and to measure the glucose levels in desired intervals of time. The nanorobot can be programmed to activate sensors and measure regularly the BGLs early in the morning, before the expected breakfast time. Levels are measured again each 2 hours after the planned lunchtime. The same procedures can be programmed for other meals through the day times. A multiplicity of blood borne nanorobots will allow glucose monitoring not just at a single site but also in many different locations simultaneously throughout the body, thus permitting the physician to assemble a whole-body map of serum glucose concentrations.

Examination of time series data from many locations allows precise measurement of the rate of change of glucose concentration in the blood that is passing through specific organs, tissues, capillary beds, and specific vessels. This will have diagnostic utility in detecting anomalous glucose uptake rates which may assist in determining which tissues may have suffered diabetes-related damage, and to what extent. Other onboard sensors can measure and report diagnostically relevant observations such as patient blood pressure, early signs of tissue gangrene, or changes in local metabolism that might be associated with early-stage cancer. Whole-body time series data collected during various patient activities levels could have additional diagnostic value in assessing the course and extent of disease.

This important data may help doctors and specialists to supervise and improve the patient medication and daily diet. This process using nanorobots may be more convenient and safe for making feasible an automatic system for data collection and patient monitoring. It may also avoid eventually infections due the daily small cuts to collect blood samples, possibly loss of data, and even avoid patients in a busy week to forget doing some of their glucose sampling. These Recent developments on nanobioelectronics show how to integrate system devices and cellular phones to achieve a better control of glucose levels for patients with diabetes.

Respirocyte - An Artificial Oxygen Carrier Nanorobot

The artificial mechanical red cell, "Respirocyte" is an imaginary nanorobot, floats along in the blood stream. These atoms are mostly carbon atoms arranged as diamond in a porous lattice structure inside the spherical shell. The Respirocyte is essentially a tiny pressure tank that can be pumped full of oxygen (O_2) and carbon dioxide (CO_2) molecules. Later on, these gases can be released from the tiny tank in a controlled manner. The gases are stored onboard at pressures up to about 1000 atmospheres. Respirocyte can be rendered completely nonflammable by constructing the device internally of sapphire, a flameproof material with chemical and mechanical properties otherwise similar to diamond.

There are also gas concentration sensors on the outside of each device. When the nanorobot passes through the lung capillaries, O_2 partial pressure is high and CO_2 partial pressure is low, so the onboard computer tells the sorting rotors to load the tanks with oxygen and to dump the CO_2. When the device later finds itself in the oxygen-starved peripheral tissues, the sensor readings are reversed. That is, CO_2 partial pressure is relatively high and O_2 partial pressure relatively low, so the onboard computer commands the sorting rotors to release O_2 and to absorb CO_2.Respirocytes mimic the action of the natural hemoglobin-filled red blood cells. But a Respirocyte can deliver 236 times more oxygen per unit volume than a natural red cell.

This nanorobot is far more efficient than biology, mainly because its diamondoid construction permits a much higher operating pressure. So the injection of a 5 cm^3 dose of 50 Percent Respirocyte aqueous suspension into the bloodstream can exactly replace the entire O_2 and CO_2 carrying capacity of the patient's entire 5,400 cm^3 of blood. Respirocyte will have pressure sensors to receive acoustic signals from the doctor, who will use an ultrasound-like transmitter device to give the Respirocyte commands to modify their behaviour while they are still inside the patient's body.

Artificial Phagocytes – Microbivores Nanorobots

A microbivore has been described, whose primary function is to destroy microbiological pathogens found in the human bloodstream, using the "digest and discharge" protocol. Nanorobotic artificial hypothetical phagocytes called "microbivores" could patrol the bloodstream, seeking out and digesting unwanted pathogens including bacteria, viruses, or fungi. Microbivores when given intravenously would achieve complete clearance of even the most severe septicemic infections in hours or less. This is far better than the weeks or months needed for antibiotic-assisted natural phagocytic defences. The nanorobots do not increase the risk of sepsis or septic shock because the pathogens are completely digested into harmless simple sugars, monoresidue amino acids, mononucleotides, free fatty acids and glycerol, which are the biologically inactive effluents from the nanorobot.

A Hypothetical Mobile Cell-Repair Nanorobot

Another nanorobot, the Chromallocyte would replace entire chromosomes in individual cells thus reversing the effects of genetic disease and other accumulated damage to our genes, preventing aging. Chromallocyte is a hypothetical mobile cell-repair nanorobot capable of limited vascular surface travel into the capillary bed of the targeted tissue or organ, followed by extravasation, histonatation, cytopenetration, and complete chromatin replacement in the nucleus of one target cell, and ending with a return to the bloodstream and subsequent extraction of the device from the body, completing the cell repair mission." Inside a cell, a repair machine will first size up the situation by examining the cell's contents and activity, and then take action. By working along molecule-by-molecule and structure-by-structure, repair machines will be able to repair whole cells. By working along cell-by-cell and tissue-by-tissue, they will be able to repair whole organs. By working through a person organ by organ, they will restore health. Because molecular machines will be able to build molecules and cells from scratch, they will be able to repair even cells damaged to the point of complete inactivity.

Further Applications of Nanorobots

Nanorobots could be used to maintain tissue oxygenation in the absence of respiration, repair and recondition the human vascular tree eliminating heart disease and stroke damage, perform complex nanosurgery on individual cells, and instantly staunch bleeding after traumatic injury. Monitoring nutrient concentrations in the human body is a possible application of nanorobots in medicine. One of interesting nanorobot utilization is also to assist inflammatory cells in leaving blood vessels to repair injured tissues.

Nanorobots might be used as well to seek and break kidney stones. Nanorobots could also be used to process specific chemical reactions in the human body as ancillary devices for injured organs. Nanorobots equipped with nanosensors could be developed to deliver anti-HIV drugs. Another important capability of medical nanorobots will be the ability to locate stenosed blood vessels, particularly in the coronary circulation, and treat them mechanically, chemically, or pharmacologically.

To cure skin diseases, a cream containing nanorobots may be used. It could remove the right amount of dead skin, remove excess oils, add missing oils, apply the right amounts of natural moisturizing compounds, and even achieve the elusive goal of 'deep pore cleaning' by actually reaching down into pores and cleaning them out. The cream could be a smart material with smooth-on, peel-off convenience.

A mouthwash full of smart nanomachines could identify and destroy pathogenic bacteria while allowing the harmless flora of the mouth to flourish in a healthy ecosystem. Further, the devices would identify particles of food, plaque, or tartar, and lift them from teeth to be rinsed away. Being suspended in liquid and able to swim about, devices would be able to reach surfaces beyond reach of toothbrush bristles or the fibres of floss. As short-lifetime medical nanodevices, they could be built to last only a few minutes in the body before falling apart into materials of the sort found in foods.

Medical nanodevices could augment the immune system by finding and disabling unwanted bacteria and viruses. When an invader is identified, it can be punctured, letting its contents spill out and ending its effectiveness. If the contents were known to be hazardous by themselves, then the immune machine could hold on to it long enough to dismantle it more completely. Devices working in the bloodstream could nibble away at arteriosclerotic deposits, widening the affected blood vessels. Cell herding devices could restore artery walls and artery linings to health, by ensuring that the right cells and supporting structures are in the right places. This would prevent most heart attacks.

Nanorobots could be used in precision treatment and cell targeted delivery, in performing nanosurgery, and in treatments for hypoxemia and respiratory illness, dentistry, bacteremic infections, physical trauma, gene therapy via chromosome replacement therapy and even biological aging. It has been suggested that a fleet of nanorobots might serve as antibodies or antiviral agents in patients with compromised immune systems, or in diseases that do not respond to more conventional measures.

There are numerous other potential medical applications, including repair of damaged tissue, unblocking of arteries affected by plaques, and perhaps the construction of complete replacement body organs. Nanoscale systems can also operate much faster than their larger counterparts because displacements are smaller; this allows mechanical and electrical events to occur in less time at a given speed.

Nanotechnology as a diagnostic and treatment tool for patients with cancer and diabetes showed how actual developments in new manufacturing technologies are enabling innovative works which may help in constructing and employing nanorobots most effectively for biomedical problems. Nanorobots applied to medicine hold a wealth of promise from eradicating disease to reversing the aging process; nanorobots are also candidates for industrial applications. The advent of molecular nanotechnology will again expand enormously the effectiveness, comfort and speed of future medical treatments while at the same time significantly reducing their risk, cost, and invasiveness.

CELLULAR BIOSCANNING

The goal of cellular bioscanning is the noninvasive and non-destructive in vivo examination of interior biological structures. One of the most common nanomedical sensor tasks is the scanning of cellular and subcellular structures. Such tasks may include localization and examination of cytoplasmic and nuclear membranes, as well as the identification and diagnostic measurement of cellular contents including organelles and other natural molecular devices, cytoskeletal structures, biochemical composition and the kinetics of the cytoplasm.

The precision and speed of medical nanodevices is so great that they can provide a surfeit of detailed diagnostic information well beyond that which is normally needed in classical medicine for a complete analysis of somatic status. Except in the most subtle cases a malfunction in any one component of the cellular machinery normally provokes a cascade of pathological observables in many other subsystems. Detection of any one of these cascade observables, if sufficiently unique and well-defined, may provide adequate diagnostic information to plan a proper reparative procedure.

DNA is one of the few cellular components that is regularly inspected and repaired. Most homeostatic systems adopt a more simple philosophy of periodic replacement of components regardless of functionality. Nanomedicine allows the philosophy of inspection and repair to be extended to all cellular components. The following discussion briefly describes a few of the sensory techniques that may be useful in achieving this objective.

CELLULAR TOPOGRAPHICS

Tactile topographic scanning provides the most direct means for examining cellular structures in vivo. Ex vivo live cell scanning in air by commercially available atomic force microscopy using a 20-40 nm radius tip allows nondestructive feature resolutions of ~50 nm across the top 10 nm of the cell. Investigators have used the scanning tip to punch a hole in the cell, then pull back and scan the breach, observing the membrane heal itself via self-assembly in real time. Up to ~KHz scanning frequencies are possible, with up to 1024 data points per scan line.

Assuming a scan rate of ~10^6 pixels/sec, a micron-scale nanodevice, once securely anchored to the surface of a human cell, could employ a tactile scanning probe to image the 0.1 Percent of plasma membrane lying within its vicinity in ~2 sec to ~1 nm^2 resolution (~1 mm/sec tip velocity), or ~50 sec to ~0.2 nm resolution. Inside the cell, and again post-anchoring, an entire 6 micron mitochondrial surface could be imaged to atomic resolution in ~100 sec; the surface of a 100-nm length of 25-nm diameter microtubule could be atomically resolved in 0.2 sec – though of course in a living cell these structures may be changing dynamically during the scanning process. Continuous power dissipation of a scan head moving at 1 mm/sec through water is ~0.002 pW (~kT/pixel at ~10^6 pixels/sec), though of course the energy cost of sensing, recording, and processing each pixel must be at least ~10 kT/pixel, so the total scanner power draw could be as high as 0.02-0.1 pW in continuous operation. Special scanning tips and techniques should allow topographic, roughness, elastic, adhesive, chemical, electrostatic, conductance, capacitance, magnetic, or thermal surface properties to be measured.

For large cellular components, identification and preliminary diagnosis of improper structure may be possible using measured surface characteristics which may be matched to entries in an extensive onboard library, perhaps combined with dynamic monitoring of anomalies in cross-membrane molecular traffic using chemical nanosensors. Most membranes are self-sealing, so it should also be possible to gently insert a telescoping member into the target organelle, which member then slowly reticulates and extrudes smaller probes with sensory tips in a fixed pattern and step size, allowing the acquisition of detailed internal structural and compositional information.

Small protein-based components of the cell such as enzymes, MHC carriers, and ribosomes are self-assembling or require only modest assistance to self-assemble. Reversible mechanical denaturation of these smaller proteins, using diamondoid probe structure to displace water molecules and by using specialized handling tools akin to functionalized AFM tips and molecular clamps, may be followed by precise nondestructive amino acid sequencing to allow identification and diagnostic compositional analysis. Afterward, the protein molecule may be refolded back into its original minimum-energy conformation, possibly with the assistance of chaperone-like structures in some cases.

Transcellular Acoustic Microscopy

Acoustic microscopy is another noninvasive scanning technique that can provide nanomedically useful spatial resolutions. Frequencies are in the GHz range, far exceeding the relaxation time of the protoplasm. A cryogenic acoustic microscope operated at 8 GHz has demonstrated 20 nm lateral resolution in liquid helium. By 1998, the best resolution achieved with water as the coupling fluid has been 240 nm at a frequency of 4.4 GHz. Operated in water at 310 K, a nanomechanical 1.5 GHz acoustic microscope would achieve a far-field minimum lateral resolution $\lambda/2 \sim 500$ nm, $\sim 10^5$ voxels per human cell, sufficient to locate and count all major organelles and a few intermediate-scale structures. Scanning acoustic microscopes operated at 2 GHz in reflection mode achieve 30-50 nm resolution in the direction of the acoustical axis, and the Subtraction SAM approach reveals topographical deviations of 7.5 nm at 1 GHz. It has been proposed that picosecond ultrasonics could be used to obtain an image of the cytoskeleton with detail comparable to that of conventional X-ray images of a human skeleton.

Power requirements are a significant constraint on acoustic reflection microscopy. In the simplest case, consider an acoustic emitter of radius r_E which converts an input power of P_{in} into acoustic power of intensity $I_E \leq I_{max}$=1000 watts/m^2 with efficiency e Percent, conservatively taken here to be 0.50. The emitter produces a train of omnidirectional pressure pulses of amplitude $A_p=(2\, r\, v_{sound}\, I_E)^{1/2}$ (N/m^2) and frequency n which travels a distance X_{path} to a target of radius r_T. As the signal reaches the target, its amplitude has been reduced by (r_E / X_{path}) due to the $1/r^2$ dependence of intensity in spherical waves and by $e^{-\alpha tiss\ n\ Xpath}$ by attenuation with $\alpha_{tiss}=8.3 \times 10^{-6}$ sec/m for soft tissue.

Upon reaching the target, a fraction $f_{reflect}$ of the signal is reflected as a point source echo; if the target surface has an acoustic impedance similar to liver tissue and the medium is similar to water, $f_{reflect} \sim 0.05$. The echo then travels a distance X_{path} back to a receiver (which may or may not be located near the emitter), the amplitude again losing (r_T / X_{path}) by geometry and $e^{-atiss\ n\ Xpath}$ by attenuation. The echo is finally detected by the receiver which can measure a pulse of minimum pressure amplitude $DP_{min} \sim 10^{-6}$ atm.

Combining these relations gives the following result:

$$P_{in} \geq \frac{\pi \Delta P_{min} X^4{}_{path}}{2 \rho v_{sound} e\% f^2{}_{reflece} r_T^2 e^{-1\alpha_{max} \nu X_{path}}} \text{(watts)}$$

To scan the entire interior of a cell, whether from within or without, requires X_{path} >~ 20 microns. A frequency of n=1.5 GHz allows v_{sound}/n ~ 1 micron spatial resolution. The fastest possible pulse repetition time is X_{path}/v_{sound} ~ 13 nanosec. Assuming r_E=r_T=0.5 micron, P_{in} ~ 7 pW or ~5 watts/m^2, well within the safe intensity range for acoustic radiation. At maximum safe transmitter intensity I_{max}, the longest scannable path length at 1.5 GHz is X_{path} ~ 46 microns.

Power constraints are somewhat less severe in the case of acoustic transmission microscopy, a technique which requires a minimum of two physically separated components. Consider an acoustic emitter radiating a series of omnidirectional pressure pulses of frequency n which travel across a tissue cell of width X_{path}, to be detected by a receiver on the other side after a signal transit time t_{cell} ~ X_{path} / v_{cell}, where v_{cell} is the speed of sound in the cytosol. In the simplest case, assume that the cytosol is clear except for one target of interest along the path in which the speed of sound is v_{target} # v_{cell}. The minimum detectable difference in travel times between a signal that passes through the target and one that does not is Δt, so the minimum resolvable target size is r_{min}~v_{target} v_{cell} Δt / 2 abs(v_{target}-v_{cell}) when r_{min} >~ v_{cell} / n. For soft tissue targets in water at 1.5 GHz, r_{min} ~ 30 microns; for tooth enamel in water, r_{min} ~ 1 micron. However, the use of sampling gates will permit the detection of phase shifts that are much less than the characteristic time constant of the system in incoming waves, hence the minimum thickness of objects in theory detectable by acoustic tomography can be far smaller than the values for r_{min} estimated above.

Computing acoustic tomographic power requirement P_{in} in a similar manner as for echolocation, except that the transmitted signal rather than the echo is detected and assuming the signals are of approximately equal strength regardless of the path taken, gives:

$$P_{in} \geq \frac{\pi \Delta P^2{}_{min} X^4{}_{path}}{2 \rho v_{cell} e\%r^2{}_{reflece} r_T^2 e^{-2\alpha_{max} YX_{path}}} \text{(watts)}$$

Using the same values as the previous example, P_{in}=1 pW for X_{path}~45 microns; for X_{path}~140 microns, P_{in}=1000 pW, near I_{max}.

In either mode of operation, data gathering is accelerated by positioning more than one receiver around the target. Sensitivity is boosted by increasing emitter size, taking multiple measurements, or by illuminating the target with some large external source located elsewhere in the tissue or even outside of the organ. However, cells and body tissues are mesoscopic "junkyards" – highly heterogeneous media which may produce large numbers of nontarget scattering events, thus increasing the difficulty of extracting signal from noise. Additional complications arise due to:

1. Scattering on rough surfaces;
2. Rapid pressure variations with range in the Fresnel zone or near field of the transmitted signal;
3. Cytoplasmic viscosity inhomogeneities due to the asymmetric arrangement of cytoskeletal structure, granules, vacuoles, and the endomembrane system; and
4. Variation in cytoplasmic Young's modulus due to time-varying tensions in semirandomly-distributed cytosolic fibrillar elements which alter elasticity and thus the local speed of sound.

Magnetic Resonance Cytotomography

Electrostatic scanning is largely ineffective because of Debye-Huckel shielding. Magnetic stray field probes allow resolution of 10 nm physical features, but only for materials with substantial magnetic domains – which most biological substances lack. But nuclear magnetic resonance imaging may allow cellular tomography by creating 3-D proton (hydrogen atom) density maps. Atomic density maps of other biologically important elements with nonzero nuclear magnetic moments (including D, Li^6, B^{10}, B^{11}, C^{13}, N^{14}, N^{15}, O^{17}, F^{19}, Na^{23}, Mg^{25}, P^{31}, Cl^{35}, K^{39}, Fe^{57}, Cu^{63} and Cu^{65}) may also be compiled. Sodium imaging is already used clinically to assess brain damage in patients with strokes, epilepsy, and tumors.

In a hypothetical NMR cytotomographic nanoinstrument, a large permanent magnet is positioned near the surface of the cell or organelle to be examined. This creates a large static background magnetic field that polarizes the protons. The spatial gradient of this field establishes a unique resonant frequency, called the Larmor frequency, within each isomagnetic surface throughout the test volume. A second time-varying magnetic driver field is then scanned through the full range of resonant frequencies, exciting into resonance the protons in each isomagnetic surface, in turn. Depending on the sensor implementation chosen, each resonance detected may cause absorption of driver field energy, an increase in measured impedance, or even a return echo of magnetic energy as the excited protons relax to equilibrium in ~1 sec. The large polarizing magnet is then rotated to a new orientation, moving the isomagnetic surfaces to new positions within the test volume, and the scan is repeated. A 3-D map of the spatial proton distribution may be computed after several scan cycles.

Only those regions within a linewidth of resonance DB will generate an appreciable signal, and the expected spatial resolution (say, along the z-axis) Dz=DB / (dB/dz), where dB/dz is the spatial flux gradient of B. For a polarizing field B=1.4 tesla placed adjacent to a cell, the cross-cell gradient dB/dz ~ 7×10^4 tesla/m; placed next to a organelle, dB/dz ~ 1.4×10^6 tesla/m. Assuming an effective NMR linewidth ~ 0.002 tesla, minimum spatial resolutions are 29 nm and 1.4 nm, respectively.

Unfortunately, the smallest reliably detectable energy in the sensor element is ~ kT e^{SNR}. The energy required to flip a single proton is E_{flip}=4 π ν_L L_{proton}, where L_{proton}=5.28×10^{35} joule-sec and the Larmor resonance frequency ν_L=g_{proton} B; g_{proton} is the gyromagnetic ratio and B is the polarizing magnetic flux. However, the population difference between spin-up and spin-down nuclei in NMR is very small. For small energy differences, the Boltzmann distribution only allows a fraction E_{flip}/2 kT to be flipped before the upper and lower state populations are made equal and the absorption disappears. In order to flip N_{flip}~kT e^{SNR}/4 π ν_L L_{proton} protons, the sample volume must include at least:

$$N_{\min} - 2e^{SNR}\left(\frac{kT}{4\pi\,\nu_L L_{\text{proton}}}\right)^2 (\text{protons})$$

For T=310 K, SNR=2 and B=1.4 tesla, ν_L=60 MHz and N_{min} ~ 1.7 x 10^{11} protons.

When scanning a cell or organelle, the vast majority of the protons present are in water molecules. Water at 310 K has a proton density n_{water}=6.7×10^{28} protons/m^3, compared to n_{fat} ~ 6.4×10^{28} protons/m^3. Distinguishing fat from water thus requires a minimum sample volume of N_{min}/(n_{water} – n_{fat}) ~55 micron^3.

Proteins range from 4.4×10^{28} protons/m^3 to 7.6×10^{28} protons/m^3, average $n_{protein}$ ~ 5.6×10^{28} protons/m^3; $n_{carbohydrate}$ ~ 5.8×10^{28} protons/m^3, allowing NMR cellular tomography to resolve minimum feature sizes of 2-4 microns (560 micron^3). This resolution may include intracellular structures such as the endoplasmic reticulum, the Golgi complex (~500 micron^3), the nucleus (~270 micron^3), possibly the nucleolus (~50 micron^3), and even large localized ferritin granule concentrations.

In 1998, the minimum sample size for microcoil (~470 micron diameter) NMR scanning was ~10,000 micron^3 or ~10^{16} protons for a 1-minute measurement cycle. Individual T-cell vacuoles ~1 micron in diameter labeled with dextran-coated iron oxide particles have

Near-Field Optical Nanoimaging

Electromagnetic waves of optical wavelength l that interact with an object are diffracted into two components, called "far-field" and "near-field." The propagation of electromagnetic radiation over distances $z > \lambda$ acts as a spatial filter of finite bandwidth, resulting in the familiar diffraction-limited resolution $\sim\lambda/2$. Classical optics is concerned with this far-field regime with low spatial frequencies $< 2/l$, and conventional optical imaging will be difficult at the cellular level in vivo. (Short-wave X-rays will damage living biological cells.) Information about the high spatial frequency components of the diffracted waves is lost in the far-field regime, so information about sub-wavelength features of the object cannot be retrieved in classical microscopy.

However, for propagation over distances $z << \lambda$, far higher spatial frequencies can be detected because their amplitudes are then of the same order as the sample ($z=0$). This second diffraction component is the "near-field" evanescent waves with high spatial frequencies $>2/\lambda$. Evanescent waves are confined to subwavelength distances from the object. Thus a localized optical probe, such as a subwavelength aperture in an opaque screen, can be scanned raster fashion in this regime to generate an image with a resolution on the order of the probe size.

The original Near-field Scanning Optical Microscope surpassed the classical diffraction limit by operating an optical probe at close proximity to the object. The NSOM probe uses an aluminum-coated "light funnel" scanned over the sample. Visible light emanates from the narrow end of the light funnel and either reflects off the sample or travels through the sample into a detector, producing a visible light image of the surface with ~12 nm resolution at $\lambda=514.5$ nm provided the distance between light source and sample is very short, about 5 nm, with signal intensities up to 10^{11} photons/sec (~50 nanowatts). This represents a resolution of $\sim\lambda/40$. The near-field acoustic equivalent is found in the medical stethoscope, which exhibits a resolution of $\sim\lambda/100$.

Applications include dynamical studies at video scanning rates, low-noise high-resolution spectroscopy, and differential absorption measurements. Optical imaging of individual dye molecules has already been demonstrated, with the ability to determine the orientation and depth of each target molecule located within ~30 nm of the scanned surface. Molecules with nanometer-scale packing densities have been resolved to ~0.4-nm diameters using STMs to create photon emission maps, and laser interferometric NSOMs have produced clear optical images of dispersed oil drops on mica to ~1 nm resolution. Live specimen 250-nm optical sectioning for three-dimensional dynamic imaging has also been demonstrated.

Submicron laser emitters have been available since the late 1980s. It should be possible to use NSOM-like nanoprobes to optically scan the surfaces of cells or organelles to <~1 nm resolution, mapping their topography and spectroscopic characteristics to depths of tens of nanometers without penetrating the surface. However, a proper membrane-sealing invasive light funnel ~20 nm in diameter might not seriously disrupt some cellular or organelle membranes and thus could be inserted into the interiors of these bodies or through the cytoskeletal interstices to permit deeper volumetric scanning. The thermal conductivity of water at 310 K is 0.623 watts/m-K and the energy per 500-nm photon is 400 zJ, so for 1 micron^3 of watery tissue the maximum scan rate is ~35 micron^3/sec-K, if e^{SNR} photons are used to image each 1 nm^3 voxel with SNR=2. Thus a 1-sec volumetric optical scan of an aqueous ~1-micron^3 sample volume to 1 nm^3 resolution requires a ~1 nanowatt scanner running at ~GHz bit rates, raising sample volume temperature by ~0.03 K.

NSOM permits the determination of five of the six degrees of freedom for each molecule, lacking only the optically inactive rotation around the dipole axis. In principle, it should be possible to produce <~1 nm resolution near-field optical scans of in situ protein molecules, since with atomically precise fabrication and single lines of atoms as conductors a minimum light guide is 3 atoms wide. Given a molecular laser and adequate collimation, photons passing through folded proteins will scatter according to the molecular structure. Detection of sufficient photons comprising these scattering patterns should allow the noninvasive determination of protein structure; polarized photons provide information on chirality. Absorption and fluorescent signals will be visible from phenyl rings, tryptophan, and bound cofactors such as ATP and adenine. Positions of monoclonal antibodies on virus surfaces are now identifiable experimentally using NSOM; it should be possible to map binding sites on virus and cell surfaces using fluorescently labelled antibodies. Single molecule detection has been proposed as a tool for rapid base-sequencing of DNA.

Other optically-based cellular imaging techniques must be distinguished. Optical Coherence Tomography uses Michelson interferometry to achieve ~10 micron spatial resolutions over tissue depths of 2-3 mm in nontransparent tissue in the near infrared. However, OCT requires numerous physical components not easily implemented on a micron-size detector, femtosecond pulse shaping, and illumination power levels of ~10^7 watts/m^2 >> 100 watts/m^2 "safe" continuous limit in tissue. Bioluminescence techniques in which the light source is placed inside the tissue has a spatial resolution limited to ~10 Percent of the depth, or ~10 microns resolution at a ~100 micron depth. Coherent anti-Stokes Raman scattering already permits organelle imaging in living cells and can in theory create a point-by-point chemical map of a cell using two intersecting lasers.

Three-dimensional observation of microscopic biological nonliving structures by means of X-ray holography requires a high degree of spatial coherence and good contrast between target and surroundings. Good contrast may be achieved in the wavelength range between the K absorption edges of carbon and oxygen, the "water window" where carbon-containing biological objects absorb radiation efficiently but water is relatively transparent. A 50-micron diameter emitter has been tested that uses near-IR 5-femtosecond laser pulses impinging upon a helium gas target to create a well-collimated (<1 milliradian) beam of coherent soft X-rays at a 1 KHz repetition rate producing a brightness of 5×10^8 photons/mm^2-milliradian^2-sec, a peak X-ray intensity of $>10^{10}$ watts/m^2 on the propagation axis behind the He target.

Cell Volume Sensing

Many cellular parameters need not be measured directly in order to be detected by a medical nanodevice. Cell volume sensing is a case in point. An intracellular nanodevice can indirectly monitor changes in the volume of the cell in which it resides by one of two methods. First, measurements of the mechanical deformation of the cellular membrane, stretch-activated channels, or cytoskeletal strains and structural changes are quite sensitive to alterations in total cell volume. Second, concentration or dilution of the cytoplasmic environment through cell shrinkage or swelling leads to the activation of various volume-regulation responses which may be detected by the nanorobot. Changes in the concentration of soluble cytosolic proteins may nonspecifically affect enzyme activity via "macro-molecular crowding". Minor changes in cell volume can cause severalfold changes in ion transport. Cellular signalling entities that have been linked to the transduction and amplification of the primary volume signal include Ca^{++} transients, phosphoinositide turnover, eicosanoid metabolism, kinase/phosphatase systems such as JNK and p38, cAMP, and G-proteins. These signal amplification pathways may be monitored using chemical concentration sensors aboard the medical nanodevice, allowing the nanorobot to eavesdrop on the natural sensory channel traffic of the cell.

NONINVASIVE NEUROELECTRIC MONITORING

In many ways the neuron is the most nanomedically important class of cell in the human body. Nanomedical applications regarding neurons and the brain are addressed in Chapter 25. The following is a brief summary of noninvasive nanotechnological methods for monitoring the electrical traffic of individual neurons. Noninvasive measurement of axonal traffic within nerve bundles will require multiple sensors and greater sensitivity to compensate for shielding by the perineurium, a tight resistive sheath enclosing the bundle ~20 microns thick with resistivity ~4000 ohm-cm.

Electric Field Neurosensing

The "typical" ~20 micron human neuron discharges 5-100 times per second, moving from -60 mV potential to +30 mV potential in $\sim 10^{-3}$ sec. Thus the variation in electric field at the axonal surface is ± 4500 volts/m. Since electric field sensors can detect 100 volt/m fields up to GHz frequencies, an electric sensor attached by circumaxonal cuff or pressed against the axonal surface should readily detect each action potential discharge. By 1998 silicon-to-neuron extracellular junctions already permitted direct stimulation of individual nerve cells in vitro without killing the cells, and extracellular electrodes were commonly used to detect neuronal electrical activity noninvasively, both in vivo and in vitro. Artificial electric fields may also be employed to trigger or moderate neural signals. Cell membrane capacitance is typically ~0.01 picofarads/micron^2, and varies with the state of the health of the cell.

Magnetic Field Neurosensing

The magnetic flux density caused by a single action potential discharge is ~0.1 microtesla at the axonal surface, which may be detected by a "compass oscillator" type magnetosensor with a ~KHz maximum sampling rate.

It might be possible for artificial magnetic fields to directly influence neural transmissions. Even a static 65 millitesla field has been shown to reduce frog skin Na^+ transport by 10-30 Percent. Each neuronal discharge develops an electrical energy of ~20 picojoule ($\sim 10^{10}$ kT), far smaller than the magnetic energy stored in a B=1.4 tesla field of a permanent micromagnet traversing an L^3=(20 micron)^3 volume which from Eqn. is $B^2 L^3/2\mu_0 \sim 6000$ pJ. If properly manipulated, such a field may be sufficient to enhance, modulate, or extinguish a passing neural signal.

Neurothermal Sensing

While neurons come in many shapes and sizes, our "exemplar" ~14,000-micron^3 neuron discharging ~90 mV into an input impedance of ~500 Kohms produces ~0.2 microampere current per pulse and generates a continuous 100-300 pW of waste heat as measured experimentally. The discharge rate of 5–100 Hz can produce brief surges up to ~2000 pW during a high-frequency train, but the duty cycle of such trains is far less than 100 Percent, reducing time-averaged dissipation to the observed 100–300 pW range.

Single impulses are measured experimentally to produce 2-7 microkelvin temperature spikes in cold or room-temperature mammalian non-myelinated nerve fibres and ~23 microkelvins in non-myelinated garfish olfactory nerve fibres at an energy density ranging from 270–1670 joules/m^3–impulse from 0–20°C. In non-myelinated fibres the initial heat occurs in two temperature-dependent phases: a burst of positive heat, followed by rapid heat reabsorption. The positive heat derives from the dissipation of free energy stored in the membrane capacity, and from the decrease in entropy of the membrane dielectric with depolarization. An L~20-micron neuron in good thermal contact with an aqueous heat sink at 310 K has thermal conductance L K_t~10^{-5} watts/K, so trains of 5-100 Hz impulses lasting 1 second should raise cellular temperature by 10-30 microkelvins; up to ~200 microkelvin thermal spikes from such trains have been observed experimentally. These events are easily detectable by nanoscale thermal sensors capable of ~1 microkelvin sensitivity up to ~1 KHz. The ~microkelvin heat signature of individual impulses or very short pulse trains can probably be temporally resolved because the minimum pulse repetition time is ~10 millisec which is much longer than the thermal time constant for an L~20-micron neuron which is t_{EQ}~$L^2 C_V$ / K_t~3 millisec.

Direct Synaptic Monitoring

The synaptic cleft between the axonal presynaptic terminal and the dendritic postsynaptic membrane is 10-20 nm in most synapses, although in the vertebrate myoneural junction it may be as large as 100 nm. Contact area per bouton is ~1 micron^2, giving a total gap volume of ~10^7-10^8 nm^3. The density of acetylcholine receptors is highest in muscles along the crests and upper thirds of the junctional folds (~10,000/micron^2), and is lowest in the extrasynaptic regions (~5/micron^2). Each action potential discharge triggers the release of ~10^4–10^5 molecules of acetylcholine into the gap volume of an active neuromuscular junction (diffusion time ~1 microsec), raising c_{ligand} from near zero to ~3×10^{-4} molecules/nm^3 (~0.0005 M) in ~1 millisec, followed by near-complete hydrolyzation by acetylcholinesterase during the 1-2 millisec refractory period. Into the gap volume may easily be inserted a ~10^5 nm^3 neurotransmitter concentration sensor able to measure ~100 acetylcholine molecules in ~1 millisec, thus detecting pulses at the fastest discharge rate. A similar device could be used to precisely regulate neuro-transmitter concentration at the junction, and hence the neural signal itself, under nanodevice control. Simple electrochemical and mechanochemical artificial synapses have been demonstrated.

CELLULAR RF AND MICROWAVE OSCILLATIONS

Starting in 1968, H. Frohlich, observing that millivolt electrical potentials maintained across cell membranes ~10 nm thick give rise to huge fields ~10^7 volts/m possibly producing an electret state, theorized that membrane molecules must be highly electrically polarized and thus could interact to produce coherent surface acoustic vibrational modes in the 10-100 GHz frequency range; the longest wavelength is about twice the membrane thickness. Interestingly, this frequency span is very close to the maximum trigger/reset frequency for bioelectronic molecules. Note that (~100 mV) (1.6×10^{-19} coul) ~ 4 kT, so a membrane molecule with a single charge on either end should be reliably reoriented by a depolarization wave, coupling pressure waves and electrostatic field fluctuations. However, the direct detection of 10–100 GHz millimeter radiation by non-nanotechnological means is experimentally difficult and controversial because the tests must be performed in vivo in close proximity to an actively metabolizing cell in water – and water strongly absorbs microwaves over macroscale ranges.

Nevertheless, active cells have shown enhanced Raman anti-Stokes scattering, an effect ascribed to the converse of the Frohlich oscillations. In one study, the normalized growth rate of yeast cultures was enhanced or inhibited when irradiated by CW microwave fields of ~30 watts/m^2 of various frequencies; growth rate data spanning 62 separate runs revealed a repeatable frequency-dependent spectral fine structure with six distinct peaks of width ~10 MHz near 42 GHz. Investigations of related phenomena are ongoing and voluminous; the interested reader should peruse *Bioelectromagnetics*, the archival journal of this field.

100 GHz waves attenuate only ~1 Percent after passing through ~3 microns of soft tissue. A single electron injected into an integral membrane protein could act as an oscillating dipole, making a 300 volt/m signal 10 nm from the protein antenna with an energy transfer of ~0.004 kT per cycle; ~1000 oscillating electrons could produce a measurable field. A 20-micron diameter cell modeled as a nonuniform spherical dipole layer with transmembrane dipoles located 10 nm apart and embedded in a dissipative medium could produce 10^2–10^5 volt/m microwave fields 1–10 microns from the cell surface..

Thus, a variety of rf and microwave electromagnetic emanations may in theory be detectable both within and nearby living cells which could prove diagnostic of numerous internal states. Such states may include cytoskeletal dynamics, metabolic rates, plasmon-type excitations due to the collective motion of ions freed in chemical reactions, positional, rotational or conformational changes in biological macromolecules and membranes, internal movements of organelles and nerve traffic conduction, cellular pinocytosis, cellular reproduction events, cell membrane identity, and cell–cell interactions.

MACROSENSING

Macrosensing is the detection of global somatic states and extrasomatic states. While the treatment here is necessarily incomplete, the discussion nevertheless gives a good feel for the kinds of environmental variables that internally-situated nanodevices could sense. Not all capabilities outlined here need be available on every nanorobot, since injection of a cocktail of numerous distinct but mutually cooperative machine species allows designers to take full advantage of the benefits of functional specialization. In many cases, a given environmental variable can be measured by several different classes of sensor device. However, since these devices are microscopic it is in theory possible to operationalize almost all of the macrosensing capabilities described in one patient using just a billion devices, a total volume of ~1 mm^3 of nanorobots or ~0.1 Percent of the typical ~1 cm^3 therapeutic dose.

CYTO-AUSCULTATION

Can sounds generated by a single cell be detected, and thus be useful for diagnosis? Probably not, given that low frequency acoustic radiators are notoriously inefficient. Mitochondrion organelles of the giant amoeba *Reticulomyxa* are shuttled back and forth by 1–4 cytoplasmic dynein motors while riding on the outside of a bundle of 1-6 microtubules. Each dynein motor generates 2.6 pN of force and drives the mitochondria at up to ~10 micron/sec, developing $0.3\text{-}1.0\times10^{-16}$ watts of mechanical power within each 320-nm diameter organelle. Taking each organelle as a cylindrical acoustic radiator with a mechanical input power of $P_{in} \sim 10^{-16}$ watts at n ~ 1000 Hz, the output acoustic pressure at the organelle surface is only ~10^{-9} atm, an acoustic power intensity $I \sim 10^{-15}$ watts/m^2 which is not detectable by micron-sized nanorobots. Nevertheless, cells and intracellular elements are capable of vibrating in a dynamic manner with complex harmonics that can be altered by growth factors and by the process of carcinogenesis, so the possibility cannot be completely ruled out.

Blood Pressure and Pulse Detection

Blood pressure ranges from 0.1–0.2 atm in the arteries to as low as 0.005 atm in the veins. The systolic/diastolic differential ranges from 0.05–0.07 atm in the aorta and 0.01–0.02 atm in the pulmonary artery, falling to 0.001–0.003 atm in the microvessels, or 0.003–0.005 atm if the precapillary sphincter is dilated. In venous vessels, pulse fluctuations are 0.002-0.010 atm in the superior vena cava, 0.004–0.006 atm in the subclavian vein, ~0.004 atm in venules generally, and ~0.0005 atm in the brachial vein. There is also a ~0.05 Hz random fluctuation in the microvessels with amplitude on the order of 0.004–0.007 atm. Both blood pressure and pulse rate can be reliably monitored by a medical nanodevice virtually anywhere in the vascular system using a $(68\ nm)^3$ pressure sensor with ~0.001 atm sensitivity.

Pulse propagation through body tissue is somewhat muted due to absorption in compressible fatty membranes, but most cells lie within 1-3 cell-widths of a capillary so the cardiac acoustic signal should still be measurable using more sensitive detectors. The time-averaged interstitial pressure in subcutaneous tissue is 0.001–0.004 atm.

Arterial pulse waves (vascular oscillations) carry subtle messages about the health of internal organs and the arterial tree. The idea of using pulse waves for diagnosis has a long history dating back 2000 years in China. Waves detected by manual probing are classified using such subjective and qualitative descriptors as floating, deep, hidden, rapid, slow, moderate, feeble, replete, full, thready, faint, weak, soft, slippery, hesitant, hollow, firm, long, short, swift, running, intermittent, uneven, taut, string-tight, gigantic, or tremulous. Abnormal waves were empirically related to disease states. Wave data gathered by nanodevices could make possible a theoretically sound, quantitative system of noninvasive observation, classification, and diagnosis as a supplement to other nanomedical tools.

Respiratory Audition

The variation of mechanical pressure over a complete respiratory cycle is ~0.003 atm in the pleura, ~0.002 atm at the alveoli, detectable by nanomedical pressure sensors positioned in the vicinity of the respiratory organs. Holding a deep breath further stretches the pulmonary elastic tissue, up to 0.02 atm.

However, turbulent flows at Reynolds number $N_R > 2300$ in the trachea, main bronchus and lobar bronchus produce a whooshing noise that may be the loudest noncardiac sound in the human torso during conventional auscultation. The energy dissipation for turbulent flow in a tube is

$$P_{turb} = P_{lam} Z = 8\pi \eta_{air} v^2 LZ \text{ (watts)}$$

where P_{lam} is the dissipation for laminar flow in a long circular cylindrical tube of length L, v is the mean flow velocity, and turbulence factor $Z=0.005\ (N_R^{3/4} - (2300)^{3/4})$, a well-known empirical formula. For $\eta_{air}=1.83\times10^{-5}$ kg/m-sec for room-temperature air, $P_{turb}=0.87$ milliwatts for the trachea; $P_{turb}=0.66$ milliwatts for the main bronchus (L=0.167 m, v=4.27 m/sec and $N_R=3210$); and $P_{turb}=0.09$ milliwatts for the lobar bronchus (L=0.186 m, v=4.62 m/sec and $N_R=2390$), totalling ~1.6 milliwatts acoustic emission from a ~120 cm^3 upper tracheobroncheal volume. This is a power density of 13 watts/m^3 corresponding to a pressure of 4×10^{-5} atm assuming a 300 millisecond measurement window at the maximum respiration rate.

The amplitude of an acoustic plane wave propagating through tissue attenuates exponentially with distance due to absorption, scattering and reflection. The amplitude is given approximately by

$$A_x = A_0 e^{-\alpha F x}$$

where A_0 is the initial wave amplitude in atm, A_x is the amplitude a distance x from the source, and α is the amplitude absorption coefficient. The function F expresses the frequency dependence of the attenuation. For pure liquids, $F=F_{liq}=n^2$ (Hz^2); $\alpha_{liq}=2.5 \times 10^{-14}$ sec^2/m for water at room temperature. However, for soft tissues, $F=F_{tiss} \sim \nu$ (Hz).

Mechanical Body Noises

Many other mechanical body noises should be globally audible to properly instrumented medical nanodevices. If normal chewing motions release 1-10 milliwatts in a ~100 cm^3 oral volume with a ~1 sec jawstroke, power density is ~10 watts/m^3 or ~10^{-4} atm of tooth-crunching noise. A stomach growl registering 45 dB at 2 meters has a source power of 160 milliwatts; released from a 10 cm^3 gastric sphincter volume gives a ~2 x 10^{-6} atm acoustic wave, detectable throughout the body. Walking and running releases 20-100 joules/footfall for a 70 kg man; assuming the energy is absorbed within a ~1 cm thickness or within ~1 second by the sole of the foot, Eqn. implies an upward-moving planar compression wave of 0.4–2.0 atm, easily detectable by acoustically instrumented nanodevices body-wide. Hand-clapping generates 0.02–0.2 atm pulses, also easily detectable.

Lesser noises including ~30 millisec hiccups at (4-60)/min, intestinal and ureteral peristalsis, sloshing of liquid stomach contents, heart murmurs, a tap on the shoulder by a friend, nasal sniffling and swallowing, clicks from picking or drumming fingernails, crepitations, manustuprations and ejaculations, the rustling noise of clothing against the skin, flapping eyelids, anal towelling, bruits due to vascular lesions, dermal impact of water while showering, copulatory noises, urethral flow turbulence during urination, transmitted vibrations from musical instruments, creaking joints, and squeaking muscles can be detected locally if not globally. Implantation of significant interconnected in vivo diamondoid structures may produce increased sensitivity to internal noises, due to the extremely low acoustic absorption coefficient of diamond.

Vocalizations

Average source power for conversational speech in air is ~10 microwatts at the vocal cords (60 dB), up to ~1000 microwatts for shouting (90 dB) and as little as 0.1 microwatts (30 dB) for whispering. Vocal cord surface area ~1 cm^2, giving an acoustic intensity I ~ 0.001-10 watts/m^2. (Using the decibel notation, dB=10 $\log_{10}$ (I/I_0), where I_0 ~ 5 x 10^{-13} watts/m^2 in air, I_0 ~ 1 x 10^{-16} watts/m^2 in water.) In a planar traveling wave, pressure amplitude A_p (N/m^2) is related to power intensity I by

$$A_P = (2\rho v_{sound} I)^{1/2}$$

For water at 310 K, ρ=993.4 kg/m^3 and v_{sound}=1500 m/sec, therefore A_p=0.0005-0.05 atm for speech, detectable by nanodevices throughout the body due to minimal attenuation at audible frequencies. Other easily detectable vocalizations include whistling, humming, coughing, sneezing, rales, wheezing, expectorating, eructations, flatus, vomiting, hawking and noseblowing.

ENVIRONMENTAL SOURCES

Can in vivo nanodevices directly detect sounds emanating from the environment outside of the body, such as other people talking in the same room or a door slamming? The waves from an external acoustic source of power P_R watts travel through the air and, upon arriving at the air/skin interface a distance x_R from the source with amplitude $A_{incident}$, are transmitted through the interface with amplitude $A_{transmit}$. For a specular reflector – interface dimensions (human body ~ 2 m) > acoustic wavelength (~0.03–3.0 m for typical audible sounds in air) – with acoustic impedance Z_1 and Z_2 on either side of the interface and perpendicular incidence,

$$A_{\text{transmit}} = A_{\text{incident}}\left[1 - \text{abs}\left(\frac{Z_1 - Z_2}{Z_1 + Z_2}\right)\right]$$

Acoustic impedance, like the speed of sound, is essentially frequency-independent over the nanomedically-relevant range of ultrasonic frequencies. For Z_{air}=400 kg/m^2–sec and assuming Z_{skin} ~ 1.6×10^6 kg/m^2-sec, then $A_{transmit}=(5\times10^{-4})$ $A_{incident}$. In other words, there is ~99.95 Percent reflection from the air-skin interface, which is why coupling mediums like gels and oils are commonly employed in ultrasound imaging. If an immediately subdermal nanodevice can detect a minimum $A_{transmit}$ ~ 10^{-6} atm, then from simple geometry:

$$P_R = \frac{2\pi x_R^2 A_{\text{incident}}^2}{\rho v_{\text{sound}}}$$

For STP (1 atm, 0°C) air, ρ=1.29 kg/m^3 and v_{sound}=331 m/sec. If the minimum detectable pressure ~10^{-6} atm ~$A_{transmit}$, then at a distance of x_R=2 meters the acoustic source must have a power of ~2000 watts, far exceeding the ~1 milliwatt output of a person loudly shouting. To hear normal conversation at x_R=2 m, minimum nanodevice detector sensitivity falls to 7×10^{-11} atm requiring a subdermal pressure nanosensor ~(17 micron)^3 in size, roughly the dimensions of a single human cell; other methods may prove more efficient.

Of course, an ex vivo acoustic nanosensor may receive sound that has passed through no interface, hence may detect pressure waves ~3 orders of magnitude lower in amplitude. Assuming $A_{transmit}$=$A_{incident}$, P_R ~ 600 microwatts, so ex vivo nanorobots with a 0.3 micron^3 sensor could hear people shouting at x_R=2 meters. To hear talking (~10 microwatt source) requires a 2.4 micron^3 ex vivo pressure sensor (limit ~10^{-7} atm).

Optimally positioned and calibrated nanomedical pressure sensors could directly measure changes in the ambient barometric pressure to within ± 10^{-6} atm. Normal atmospheric variation due to weather ranges from 0.94-1.05 atm; such slow moving changes are readily monitored. Very near the Earth's surface, the air pressure P at altitude h above sea level is approximated by $P=e^{-k_p h}$ (atm), where k_p=1.16 x 10^{-4} m^{-1} at 20°C; at sea level, a 10^{-6} atm change in pressure reflects a change in altitude of only ~1 cm. However, the opening or closing of a door inside a (~5 m)^3 room that displaces >125 cm^3 of air also causes a minimally detectable >10^{-6} atm pressure pulse. Other sources of environmental pressure variation such as infrasonic (~0.2 Hz) microbaroms from offshore ocean storms, wind entering through open windows, forced-air currents from central heating or A/C systems, or even the movements of nearby people and pets may be detectable and thus may further confuse the measurement, reducing absolute accuracy unless suitable corrections are made.

KINESTHETIC MACROSENSING

Using a navigational transponder network, a population of ~10^{11} nanodevices each spaced an average ~100 microns apart throughout the body tissues can determine relative location to a positional accuracy of ~3 microns and an angular accuracy of ~2 milliradian, with the data updatable once every millisecond. Thus dispersed, the network can continuously monitor and record the relative positions of all limbs with worst-case ~0.8 mm accuracy over a 2-meter span. These devices can prepare high-resolution dynamic maps of body position, velocity, acceleration and rotation, allowing precise real-time digital kinesthesia during sports or artistic activities such as gymnastics, pole vaulting, or ballet; during transportative activities such as driving cars around hairpin turns, roller-coaster rides, military aircraft maneuvers and space launches; during precision tool-using such as needle-threading, antique watch repair, or while using the fingers as measurement calipers; during self-defence activities requiring complex motions such as karate or judo; and during emergency situations such as automobile crashes and tumbling motions during falls from great heights. All such sensory data is readily outmessaged to the patient or user in real time.

Nanorobots could assist in path integration or the reconstruction of experienced limb trajectories. Given the ability to determine rotation rates and to measure applied forces and torques, the network should also be able to forecast the anticipated future positions of limbs. By comparing these projections to actual results, the network can then infer the viscosity of the medium in which the activity is taking place, whether the environment is stable or is translating or rotating in some direction, and whether the human user is physically supported or in free-fall. Internal nanodevices can directly measure if a body is sitting, standing, laying prone, inverted, falling, floating or diving, and then communicate that information directly to the human user. Detection of patient activity states – e.g., the patient is sitting, standing, or walking – can be used to control nanorobot behaviours, activate or deactivate outmessaging displays, and so forth. A simple macroscale wearable tactile "compass belt" that would convey desired geographical directions has already been proposed.

Orientational Macrosensing

Since nanopendular sensors can reliably determine the direction of the local gravity field vector to within ~2 milliradians in $\sim 10^{-4}$ sec, the navigational network can poll its members and arrive at an accurate consensus on which direction is up. Nanodevices affixed to relatively stable hard body parts will exhibit more consistent orientational readings. Measurement of the gravity vector allows the vertical orientation of the rest of the body to be precisely fixed in space, especially useful for gymnasts, trapeze artists, underwater divers in murky lakes, and vestibular-impaired individuals, to whom this information may be outmessaged directly.

The gravitational vector may also be indirectly measured, albeit more slowly and less accurately, by monitoring the body's natural reactions to changes in the axis of gravitational loading. The variation in head-to-toe hydrostatic pressure for a 1.7 m tall, 70 kg human standing in a 1-g gravity field is ~0.17 atm; the ability to measure a systematic cross-body differential of 10^{-6} atm allows the detection of a $\sim 10^{-5}$ g change along the lengthwise aspect, or $\sim 10^{-4}$ g change along the transverse aspect. Thus a continuously updated whole-body barostatic map allows the human body to serve as a three-dimensional gravity/orientation sensor.

Body Weight Measurement

A network of nanodevices can inventory the volume and density of each of the body's $\sim 10^{8}$ $(1\ \text{mm})^3$ voxels using a combination of acoustic ranging, somatic mapping, and flowmetry. Each voxel is identified as fat, muscle tissue, bone mass, interstitial fluid, and so forth. These measurements allow the body's weight to be precisely computed as volume times density, rather than the usual method of determining weight using gravitational force data. The existence of a natural physiological "ponderostat," a crudely analogous humoural body-mass detector, has been proposed; indeed, the insulin-leptin system has been found to serve a related purpose, and is easily eavesdropped by medical nanorobots.

Gravitational Geographic Macrosensing

Medical nanodevices can measure variations in the gravity field to ~10^{-6} g's for L=20 micron gravimeters in a measurement time t_{meas}=2–9 millisec. This implies that in vivo nanodevices can take precise measurements of their latitude and altitude relative to sea level ~100 times every second. Gravity increases toward the poles and at lower altitudes. Specifically, using the formula of Cassinis with the Bouguer correction to the free air variation by altitude, measured gravity g_{meas} is given approximately by

$$g_{meas} = g_0\left[1 + k_1 \sin^2(\theta_L) - k_2 \sin^2(2\theta_L)\right] - k_3 h + k_4 h \rho_{earth}$$

where θ_L=terrestrial latitude (equator=0°), h=height above sea level in meters, g_0=9.78039 m/sec^2 (equatorial sea-level value ofg), k_1=5.2884×10^{-3}, k_2=5.9×10^{-6}, k_3=3.086×10^{-6} sec^{-2}, k_4=4.185×10^{-7}, and ρ_{earth}=5522 kg/m^3.

Since sea-level g varies from 9.78039 m/sec^2 at the equator to 9.83217 m/sec^2 at the north pole, a 20-micron gravimeter (Δg=10^{-6} g) detects a change in position of 1 arcmin of latitude or ~1900 meters north/south along the Earth's surface. Similarly, since at 45° latitude g varies from 9.806 m/sec^2 at sea level to 9.803 m/sec^2 at 1000 meters altitude, a 20-micron gravimeter detects a change in altitude of ~3.3 meters. For comparison, in 1998 high-quality commercial gravity gradiometers measured gradients of ~10^{-9} g/meter and allowed the compilation of micro-g (~1 milligal) resolution aerial gravity maps; atom interferometers measured the gravitational acceleration of atoms to a precision of 10^{-10}.

To achieve such phenomenal positional accuracies, the nanodevice must be able to computationally resolve several complicating factors. First, localized mass concentrations representing nonuniformities in crustal density produce residuals of up to ±0.0006 m/sec^2, which may be removed from the data using a standard map of known terrestrial isostatic variations and anomalies. Indeed, matching observations to such a map could provide useful longitudinal information as well. Another complication is the variation in gravity due to tidal forces amounting to $\sim 3\times10^{-7}$ g's, twice daily, which lies at the limits of detectability for a 20-micron gravimeter. Other minor geodesic and terrain-related corrections, too complicated for discussion here, may also need to be applied in certain circumstances. Note that the presence of nearby heavy objects does not influence measurement accuracy: a 100-ton building 10 meters away adds a lateral acceleration of only 7×10^{-9} g's to a human body.

One final complication is that patient movements create kinematic accelerations that must be distinguished from the gravitational accelerations. Gravity readings can be corrected by taking derivatives of the signals from kinesthetic monitoring to give gravity in the reference frame of the patient's room, and many individual measurements may be averaged to improve accuracy since the gravity vector normally changes only very slowly.

VASCULAR-INTERSTITIAL CLOSED ELECTRIC CIRCUITS

In vivo studies of the electric properties of blood vessels shows that the walls of veins and arteries present a specific electric resistance ~200 times greater than the conductive medium of blood — roughly 200 ohm-m vs. 0.7 ohm-m. Thus blood vessels are properly regarded as relatively insulated conducting cables which can electrically connect an injured tissue with surrounding noninjured tissue. Capillaries form an electric junction with the interstitial fluid, so the electric gradient can be canceled by ionic transport. Cell membranes are insulating dielectrics with resistance and capacitance, penetrated by ionic channels or gates for ionic transport. Additionally, in 1941 Szent-Gyorgyi suggested the possibility of semi-conduction in proteins, a theory which has since been elaborated by many investigators.

B. Nordenstrom has proposed that the above schema represents a system for selective electrogenous mass transport of material between blood and tissue which he calls Vascular-Interstitial Closed Circuits — in effect, an additional electro-circulatory system operating in parallel with the well-known diffusive, osmotic, and hydrodynamic mechanisms of regular bloodstream materials transport. As long as ions can leak through open pores and migrate through ion channels, long distance transport cannot take place and the VICC system remains primarily a local, not global, electrical circuit. Connections also exist with conductive media including cerebrospinal fluid, bile, and urine.

Nanorobots capable of monitoring the status of local VICCs may rapidly and efficiently acquire a wealth of systemic information without the need for direct inspection of the affected tissues. When a working muscle produces catabolic products such as lactic acid, an electrochemical potential gradient develops between the muscle and surrounding tissue in a known manner. Injured tissues become polarized in relation to surrounding tissue, as initial catabolic degradation products acidify the tissue. Malignant neoplasms, benign neoplasms, and internally necrotic granulomas all may polarize electrically in relation to surrounding tissue; a vascular thrombus also contains ionized material. Large deflections in the diffusion potential of blood occur if blood is deoxygenated or is infected with Gram-negative bacteria. Blood changes its electric potential from +500 mV to +1000 mV during spontaneous coagulation, and the spontaneous electric potential at the site of a crush injury of the iliac crest in rats oscillated several times between +190 mV and –60 mV over a four day experimental trial.

Nordenstrom proposes that in vivo electrophoresis via the VICC system may also play a role in mediating leukotaxis. When tissue is artificially polarized by electrodes, accumulation of granulocytes with margination and development of pseudopods is the result. In one experiment, exposure of a mesenterial membrane to a 1 microampere 1 volt field for 30 minutes stimulated diapedetic bleeding near the anode. At higher power, the arterio-capillaries contracted, narrowed, and emptied of blood cells, while the veno-capillaries and venules widened and filled with granulocytes. Monitoring such actions of local VICCs experiencing natural voltage fluctuations could alert medical nanorobots to the presence of local injuries, tissue changes, or pathologies that otherwise might go unnoticed for a considerable time.

Electric/Magnetic Geographic Macrosensing

Detection of electric fields to ~100 volts/m, the normal vertical atmospheric gradient, in theory might allow determination of aboveground altitude to ~1 m accuracy if the time between recalibrations is very brief. However, a human body which is a good conductor, standing on level ground, acquires a slight negative surface charge and becomes "part of the ground" electrically. This distorts upward the equipotentials that usually run parallel to the ground, thoroughly corrupting an altitude measurement. Nearby lightning hits and storms are detectable, which is useful.

However, medical nanodevices with appropriate magnetometers can measure variations in magnetic field to ~0.1 microtesla using a (660 nm)^3 permanent magnet sensor, in a measurement time t_{meas} ~ 0.6 millisec. Geomagnetic field maps of the Earth's surface are highly nonisotropic in both latitudinal and longitudinal directions; knowledge of the field plus the absolute direction of true north gives longitude information from the separation of the planetary magnetic and spin poles. This implies that properly equipped in vivo nanodevices can establish their latitude and longitude on Earth's surface ~1000 times every second by this means. In particular, the horizontal component of the geomagnetic field vector ranges from 0-41 microtesla from magnetic pole to magnetic equator; the independent vertical field component ranges from 0-70 microtesla. Thus a 0.1-microtesla sensor should resolve each component to an accuracy of ~11 arcmin or ~20 km in both latitude or longitude. Terrestrial surface magnetic anomalies produce local variations of 0.03–30 microtesla and are thus detectable. Additional complications to these measurements include artificial magnetic field sources, the 0.01–0.1 microtesla solar daily variation, occasional magnetic storms producing erratic geomagnetic fluctuations of 0.01–5 microtesla, sudden-commencement ionospheric electrojets up to 0.3 microtesla, and various long-term secular variations in the geomagnetic field.

Radio, television, and direct broadcast satellite signals probably are not detectable by individual nanodevices, but such detection may be indirectly achieved in vivo using nanorobot-accessible dedicated macroscale antennas implanted in the body.

Piezoelectric Stress Macrosensing

The piezoelectric effect is "the production of electrical polarization in a material by the application of mechanical stress". Many materials in the human body are piezoelectric, including tendon and elastin, dentin and bone. This polarization, or surface charge, varies as the physical stress imposed on the material changes over time. Thus, measurement of the piezoelectric effect in bone or tendon can provide information on the physical loads that are being carried by these materials. A shear stress applied along the long axis of a bone alters the polarization voltage that appears on a surface at right angles to the axis. In another experiment, piezoelectric surface charges on a loaded human femur were measured to range from -131 to +207 picocoulombs/cm^2 (-8 to +13 charges/$micron^2$) from one end of the bone to the other end, depending upon position along the shaft. Real-time monitoring of this data would permit specific inferences as to the amount of load the bone was carrying and from what direction, including bending, shearing, and twisting forces, from which whole-body activity states could subsequently be inferred. Knowledge of these surface charge variations might also be exploited in osteographic or functional navigation.

Optical Macrosensing

In vivo medical nanodevices can gather little useful optical information from the external environment. Here's why.

In biological soft tissues, scattering dominates absorption except in the pigmented layers of the epidermis and stratum corneum. Thus the propagation of light in tissues may be regarded as occurring in two steps.

In the first step, optical photons of intensity I_0 falling perpendicularly on skin are transmitted according to Beer's law through tissue to a depth z with a transmitted intensity of

$$I_x = I_o\left(1 - r_{xp}\right)e^{\alpha_t x}\left(watts/m^2\right)$$

where r_{sp}=specular reflection coefficient for visible light ~ 4 Percent-7 Percent, or 0 Percent if the original light source lies within the body; and the transmission coefficient σ_t=σ_a (absorption coefficient ~ 300 m^{-1}) + σ_s (scattering coefficient ~30,000 m^{-1}) for various human soft tissues at optical wavelengths. (Coefficient values for the most heavily pigmented skin layers may be 5-7 times higher.) Typical values for σ_t ~ 10,000-100,000 m^{-1}, average ~30,000 m^{-1} for soft tissue, although exceptionally clear tissues with σ_t=1000 m^{-1} have been reported. Thus the mean free path of an optical photon in human soft tissue is 10-100 microns, average ~ 30 microns (~1.5 tissue cell-widths), up to an extreme maximum of ~1 mm for the most transparent tissues known. For σ_t ~ 30,000 m^{-1}, at z=150 microns, I_z / I_0 ~ 0.01 and ~99 Percent of all photons have been scattered at least once from their initial path.

In the second step, in an unbounded medium the patch of fully scattered photons continues to propagate through the tissue via diffusion until all photons are absorbed. The complicated governing diffusion equation has not yet been been completely solved analytically, but the asymptotic diffusive fluence is given roughly by

$$I_d - I_i e^{-\alpha_d x}\left(watts/m^2\right)$$

where σ_d is the diffusion exponent or effective attenuation coefficient (averaging ~900 m^{-1} for typical soft tissues but with an extremely wide range reported over the ultraviolet, visible, and near-infrared wavelengths, from 10-1,000,000 m^{-1}.) As a crude approximation, the initial intensity of the fully scattered photon patch I_i ~ I_0 (1-r_{sp}) a_{scat}, where a_{scat}=σ_s / (σ_s+σ_a) ~ 0.987 is the albedo for single particle scattering.

Over the optical band from 400-700 nm, incident intensity I_0=100-400 watts/m^2 when standing in direct sunlight; artificial lighting in homes and offices is typically 0.1-10 watts/m^2; moonlight provides only I_0~10^{-4} watts/m^2; and the absolute threshold for human vision is ~10^{-8} watts/m^2. For σ_d~900 m^{-1}, the intensity ratio of transmitted/incident visible light falls to I_d/I_0=0.1 at z=2.5 mm depth (~eyelids), ~10^{-4} at z=1 cm depth, and ~10^{-11} at z=2.8 cm – a depth at which the tissue would be completely dark to the human eye even under direct sunlight illumination at the outermost skin surface.

An optical nanosensor with N_{sensor} receiver elements, with each receiver element having an area A_e=1 nm^2 and capable of single-photon detection, requires an integration time for reliable detection of e^{SNR} photons given by

$$t_{meas} = \frac{h\nu e^{SNR}}{I_d N_{sensor} A_e}(\text{sec})$$

where h=6.63 x 10^{-34} joule-sec and n=4.3-7.5×10^{14} Hz for optical photons. Let us assume that a patient is standing in I_0 ~ 400 watt/m^2 direct unfiltered sunlight, and that t_{meas}=1 sec, SNR=2, N_{sensor}=25,000 elements giving a nanorobot eyespot of A_e N_{sensor}=0.025 micron^2, and that s_d=900 m^{-1}. Then the maximum tissue depth for optical photon detection is z_{max} ~ 17 mm (I_d ~ 10^{-4} watt/m^2) which includes a volume of tissue comprising up to ~30 Percent of total body volume. Changes in illumination at the minimum indoor artificial level of ~0.1 watts/m^2 are visible to z_{max} ~ 7 mm depth. This eyespot, if exposed to air on the outermost surface of the skin, is just sensitive enough to detect full moonlight.

Our general conclusion is that variations in normal indoor lighting may be directly measurable by nanorobots stationed within the outermost ~1 cm of body tissues. Imaging, as opposed to the simple illumination detection system described here, is a much more difficult design challenge.

NEURAL MACROSENSING

The ability to detect individual neural cell electrical discharges noninvasively in many different ways, coupled with the abilities (A) to recognize and identify specific desired target nerve cells and (B) to pool data gathered independently by spatially separated nanodevices in real time, offers the possibility of indirect neural macrosensing of complex environmental stimuli by eavesdropping on the body's own regular sensory signal traffic.

Nanomonitors positioned at the afferent nerve endings emanating from hair cells located in the otolithic membrane of the utricle and saccule and in the cristae ampullaris of the semicircular canals allow medical nanodevices to directly record, amplify, attenuate, or modulate the body's own sensations of gravity, rotation, and acceleration, although kinesthetic sensory management might also be required for complete control. Motor neurons likewise can be monitored to keep track of limb motions and positions, or specific muscle activities, and even to exert control. Neuron-resident nanorobots may detect auditory effects induced in the brain by pulsed microwaves from external sources. Feline cochlear neurons sensitive to audible frequencies of >300 Hz respond to single microwave pulses at a threshold-specific absorption rate of 6-11 watts/kg-pulse, while in humans the threshold for effect, which depends on energy per pulse, may be as low as ~0.02 joule/m^2-pulse for people with low hearing threshold.

Olfactory and gustatory sensory neural traffic similarly may be eavesdropped by nanosensory instruments. Nerve taps in the medulla oblongata or at the phrenic nerve that drives the diaphragmatic muscles allow direct monitoring of respiratory activity. Pain signals may be recorded or modified as required, as can mechanical and temperature nerve impulses from other receptors located in the skin. Even psychological variables such as emotionality, vigilance, and mental workload may be directly monitored by measuring ANS activity in sympathetic efferent fibres outside of the brain.

The most complex and difficult challenge in neural macrosensing will be optic nerve taps. The retina is thoroughly vascularized, permitting ready access to both photoreceptor and integrator neurons. However, the optic nerve bundle itself has ~10^6 tightly bunched individual nerve fibres, a 10-100 MHz signal bandwidth, and significant natural data compression techniques which all must be untangled in real time.

Developing algorithms capable of interpreting raw optical nerve traffic, say, to recognize a specific human face or a specific scene in the vision field, would prove a significant research challenge. Direct monitoring of photo-receptors or retinal membrane potentials and other techniques may simplify untangling of the signal compression. Rapid visual field identification using artificial neural nets is also a subject of much current research interest. Eyeball rotations, eyelid position, pupil aperture, and lens accommodation under ciliary muscle control must be monitored to supplement the vision field analysis.

7

Navigation

NAVIGATING THE HUMAN BODY

It is difficult to imagine any significant application of medical nanodevices which does not involve navigation, however crude. Devices intended to monitor somatic states, assemble artificial internal structures, remove tumors or foreign matter, combat infections, or perform repairs, must normally be extremely tissue- or cell-specific. Navigation is also required to execute many control protocols, to locate dedicated energy, communication, or navigational helper organs, or to stationkeep and coordinate with other nanodevices. Even bloodborne nanorobots intended to operate solely at the systemic level — such as nanobiotics or immunocytes and respirocytes — must know if they have been prematurely ejected from the vasculature so that they may cease functioning or at least modify their activities.

Perhaps the most important challenge of in vivo navigation is to determine how physicians may best direct nanorobots to specific target sites needing treatment within the human body. Two alternative strategies appear most likely to produce the best clinical results.

The first strategy is positional navigation, in which the nanorobot knows its position inside the human body to ~micron accuracy at all times in some clinic-centered or body-centered coordinate grid system. The nanodevice relies upon dead reckoning, cartotaxis, microtransponder network alignment, or triangulation on external beacon signals to establish its position continuously. This method requires some onboard computation, at least a basic set of sensors, and probably also a good clock. However, it is hardly foolproof – if the target coordinates are poorly specified or the beacon signals are misaligned or poorly calibrated, the nanorobots may go to the wrong place.

The second strategy is functional navigation, in which nanodevices seek to detect subtle variations in their environment, comparing diverse sensor readings with the profile of the target tissue or cell and congregating wherever this very precisely defined set of preconditions exists. These preconditions may be thermal, acoustic or barostatic, cytochemical or immunochemical, mechanical or topological, or even genetic. The crudest forms of functional navigation may be called demarcation, wherein the doctor manually creates detectable artificial conditions at or near the target site such as dermal hot spots, injected chemical plumes, or focused ultrasonic beam spots of appropriate magnitudes and frequencies. Demarcation strategies can be implemented using extremely simple onboard sensors and control devices, possibly not even requiring a nanorobot computer, thus may prove useful early in nanomedical technology development.

More sophisticated forms of functional navigation can be extraordinarily flexible because targets may be specified without the physician having to know their exact physical location in the body – e.g., nascent cancer tumors, T cells reactive to specific antigens, infected deep-thoracic lymph nodes, bacteria of a particular species, broken capillary vessels, or virus particles having a specified protein coat chemistry. The physician need not know the exact number of targets, nor even if any targets are present at all. These forms of functional navigation require onboard computation, a resident database of relevant parameters and operational details, and a wider assortment of sensors and control protocols. But they also offer the greatest benefit at lowest risk for the patient and thus must be regarded as the preferred approach once the technology is available.

This Chapter opens with a survey of human somatography, followed by general discussions of positional and functional navigation, cytonavigation, and finally ex vivo navigation.

Human Somatography

Somatography is, quite simply, the "geography" or map of the anatomical spaces, as seen from the viewpoint of a microscopic traveller in those realms. The human body is a complex and fascinating place to visit. The nanomedical theater of operations in adult patients may range in size from an extremely small ~0.005 m^3 to an extremely large ~0.5 m^3 but averages ~0.06 m^3 of navigable volume for the standard 70 kg adult male having 15 Percent body fat, medium build and good health. Thus the volume normally available for nanorobot navigation in a single patient is ~10^{17} $microns^3$.

The first fully digitized comprehensive three-dimensional map of the human body was compiled in the early 1990s as part of the National Library of Medicine's Visible Human Project. During this effort, a male and female cadaver were frozen in blocks of gel, sectioned into thousands of thin slices and digitally scanned in MRI (magnetic resonance imaging), CT (computerized X-ray tomography) and anatomical modes (optical photography), slice by slice.

The male data set, released in 1994, consists of axial MRI images of the head and neck taken at 4 mm intervals and longitudinal sections of the rest of the body also at 4 mm intervals; each MRI image is 256×256 pixels with a 12-bit grey scale per pixel. The CT data consists of 512×512 pixel axial CT scans of the entire body taken at 1 mm intervals, also with a 12-bit grey scale per pixel. The axial anatomical images are 2048×1216 pixels with a 24-bit colour scale per pixel, representing ~60 megabits per image. The anatomical cross-sections are also at 1 mm intervals and coincide with the CT axial images. There are 1871 cross-sections for each mode obtained from the male cadaver, a ~0.112 terabit data set for the anatomical images.

The female data set, released in 1995, has the same characteristics as the male set except that the axial anatomical images were obtained at 0.33 mm intervals instead of 1.0 mm intervals, resulting in more than 5,000 anatomical images and a data set of ~0.336 terabits. Even though this map has a resolution of only 37 million micron^3 per voxel, this is still sufficient to resolve all terminal veins and terminal arterial branches in the body and all major gross anatomical features, and thus provides a good start toward a human somatographic atlas.

A complete human somatographic static map to cellular resolution in theory requires ~20 micron increments, a ~100 terabit data set using 8-bit 8000 micron^3 voxels assuming a fixed scanning geometry and ignoring the indexing tables. Recording all major structural details in the capillary terminal bed demands ~4 micron resolution, a ~10,000 terabit data set assuming 8-bit voxels. A 10,000 terabit data set may be stored on a ~26 bit/nm^3 hydrofluorocarbon memory tape in a cubic volume of ~(72 micron)^3 within an in vivo ~(100 micron)^3 library nodule. Divided into ~10^6 independent spools, a datum located anywhere on the 3300-kilometre total tape length could be accessed in ~10 seconds assuming a read speed of ~30 cm/sec. A smaller 100 terabit library nodule requires only ~(16 micron)^3 of tape, with ~1000 spools and similar access times as in the previous example. Topological or functional mapping may permit data compression by a factor of 10-100 without increasing access time or seriously reducing map utility.

A 1 micron storage block that could conveniently be carried aboard an individual medical nanorobot can hold ~0.01 terabits, enough memory to contain a three-dimensional map of the entire human body to ~430 micron resolution or a map of a 1-kg organ to ~100 micron resolution, using 8-bit voxels. The required resolution for a given application is very missicn-dependent. Additionally, two types of map are likely to be of value. The first is a precise static map of some generic reference cadaver, as described above, that may provide general navigational guidance. The second is a detailed map of the individual patient, assembled by exploratory or "surveyor" nanorobots prior to the deployment of therapeutic nanorobots which would be given this information to allow very specific navigational guidance. Map stability is an important issue. The remainder of this Section offers a quick guided tour of the larger navigable volumes inside the human body, along with some useful quantitative details. The different regions of the body appear to have enough structural and chemical dissimilarity to allow a nanorobot traversing these volumes to determine its position to at least ~mm accuracy, based solely on simple landmark recognition, chemonavigational cues, and bifurcation and topological information, even without resorting to the precision positional systems described.

Navigational Vasculography

Medical nanorobots may access the interior of the human body principally via the blood-carrying vasculature, representing ~5.4 liters or ~9 Percent of total body volume. The blood vessels are of three types, each characterized by functional and structural differences. First, there is the high-pressure distributing system, made up of the arteries and smaller branches called arterioles, which convey oxygenated blood from the heart to all regions of the body. Second, there is a network of minute vessels, called capillaries, through which biologically important materials are exchanged between blood and tissues. Third, there is a low-pressure collecting system made up of veins and smaller branches called venules, which return the blood to the heart. The lymphatic system, an additional 3.3 liters of very low-pressure vasculature, returns cell-filtered plasma to the main circulatory system.

Arteriovenous Macrocirculation

Topologically, the arteriovenous circulatory system resembles a "figure-eight" with a four-chambered heart positioned at the central junction. In the top half of the "figure-eight," oxygen-depleted blood is pumped from the heart to the lungs via the pulmonary artery. Oxygen-rich blood leaves the lungs and returns to the heart via the left and right pulmonary veins. In the bottom half of the "figure-eight," oxygen-rich blood received from the lungs is pumped into the aorta for distribution to the tissues. Oxygen-depleted blood is collected by the venous network and returns to the heart via the vena cava. Most veins 1 mm in diameter or larger are fitted with a series of one-way valves to prevent backflow. Such valves are most numerous in the lower extremities: the vena cava and the mesenteric, pulmonary and portal veins. Arteries have no valves. The smaller arteries are heavily anastomosed except across the midline of the body.

How much time does a nanorobot flowing with the blood take to transit individual organs and the entire vascular circuit? It is generally accepted that the roundtrip circulation time through the entire "figure-eight" averages ~60 seconds under resting conditions, a time that appears constant for nearly all mammals. During heavy exercise, the minimum circulation time may fall to 11-15 sec along most pathways primarily due to vasodilation.

Tissue transit times have been widely studied. Bloodborne particles in the lung normally take only 1-2 seconds to pass the alveolar sheet but 5-10 seconds to transit the entire pulmonary vasculature. Lung transit time is not significantly reduced if heart rate is simply increased; however, during strenuous exercise the alveolar transit time falls to 0.3 sec and the maximum lung transit time falls from 10 sec to 2 sec due to vasodilation. Organ transit times are dominated by capillary flow speeds which range from 0.20–1.50 mm/sec. Given that capillaries average ~1 mm in length, minimum organ transit times are 0.7–5.0 seconds.

NAVIGATIONAL BRONCHOGRAPHY

Medical nanorobots may access the human body via the respiratory system. The airway begins at the mouth and nose, extends through the pharynx, larynx, the trachea and bronchial branchings, and ends at the alveoli. The conducting zone from the top of the trachea to the beginning of the respiratory bronchioles contains no alveoli, so gas exchange with the blood does not occur there. Gas exchange occurs only in the respiratory zone, which extends from the respiratory bronchioles down to the alveolar sacs.

The respiratory airflow begins in the mouth and nose. The nasal cavities are divided by a partition called the nasal septum, from which diverge three winglike projections called the conchae. The endings of the olfactory nerve lie in the mucosa in the cilia-free region near the ~1 cm^2 olfactory bulb above the superior concha. The nasal passages are flanked by four sinuses which may swell up during infection or inflammation, closing off the air passages and necessitating breathing through the mouth. Lacrimal ducts drain tears and other secretions from the eyes into the nose via the nasolacrimal duct, requiring noseblowing after crying. The <~1 mm human vomeronasal organ, located near the base of the nasal septum in adults, transduces pheromonal signals, although apparently this organ is not present, is inactive, or is very insensitive in some people.

Air passing through the nasal cavities is warmed to within 1 K of body temperature by an extensive capillary vasculature, and is humidified by nasal mucous glands (e.g., goblet cells which secrete mucoid fluid) to within ~1 Percent of full saturation, before the air enters the pharynx. Coarse particles are removed from incoming air by nose hairs. Smaller particles are removed by turbulent precipitation, wherein obstacles in the nasal passages force the airflow to execute many sharp turns which the particles are too heavy to negotiate. The particles hit the nasal mucous membrane and become embedded in the mucus. The nasal turbulence mechanism is so effective that almost no nasally-inhaled particles larger than 2-5 microns reach the lower airway.

Most of the surfaces of the nasal airway are covered by a layer of ciliated, pseudostratified columnar epithelium cells. Each epithelial surface cell has 25-100 cilia that beat forcefully and continually toward the pharynx at ~10 Hz. Respiratory cilia are ~0.2 microns in diameter and ~2-5 microns long with a mean separation between cilia of ~2-5 microns. The film of particle-carrying nasal mucus >~200 microns thick is moved toward the pharynx by the cilia at a speed of ~1-3 cm/min, where it is then swallowed down the esophagus. Mucus velocity increases with luminal depth. In unciliated areas such as the front of the nose, where temperatures fall below levels the cilia can tolerate, the mucous layer creeps along the surface solely through traction from neighbouring ciliated areas. Swallowed mucus volume totals ~0.1 cm^3/min. The absolute viscosity of normal mucus is typically ~1 kg/m-sec, rising as high as ~500 kg/msec in the thick sputum of cystic fibrosis patients. Mucus rheology and mucociliary clearance mechanisms of the respiratory tract have been widely studied, along with the energy dissipation per cilium in the periciliary fluid. Airflow contributes little to mucus transport during normal breathing, except in patients with bronchial hypersecretions, although high-frequency ventilation and coughing may make significant contributions.

During inspiration, air passes through the nose or mouth into the pharynx (throat). The posterior wall of the pharynx rests against the cervical vertebrae; the lateral wall has openings communicating with the middle ear. The pharynx branches into two tubes — the esophagus and the larynx. The larynx houses the vocal cords, two strong bands of elastic tissue stretched horizontally across its lumen. The flow of air past the vocal cords causes them to vibrate, producing sounds. The ventricular folds or false vocal cords point downward. When closed by the action of sphincter muscles, the folds form an exit valve that permits the building up of abdominal pressure as in straining at stool or in expulsion of the fetus. Coughing involves releasing the air explosively through the folds.

After passing the larynx, air enters the trachea, a cylindrical tube with 16-20 circumferential cartilaginous rings shaped like horseshoes and embedded in an external fibroelastic membrane. The trachea branches into two main bronchi, one of which enters each lung. The right bronchus is shorter, wider, and more nearly vertical in direction than the left bronchus; both are supported by complete rings of cartilage to prevent collapse under high levels of suction. There are more than 20 generations of branchings in the lungs, each resulting in narrower, shorter, and more numerous tubes. When the bronchi become smaller than ~1 mm in diameter, they lose their cartilage, and become bronchioles.

The lungs themselves are cone-shaped organs which lie in the pleural cavities of the thorax. The base of each lung contacts with the upper surface of the diaphragm, extending to the level of the 7th rib anteriorly and the 11th rib posteriorly. The right and left pleural cavities are formed by two serous sacs into which the lungs are invaginated. Two layers of pleura are separated by a thin layer of fluid from ~20-80 microns thick. Pleural fluid is formed on the parietal pleural surface at a rate of 7-11 cm^3/hr, with up to 20-25 cm^3 normally present in the pleural space; glucose is present at serum levels and there are topographic differences in pleural pressure. The right lung, larger in size than the left, is divided into three lobes; the left lung has only two lobes presumably to make room for the heart. The interlobar surfaces are covered with visceral pleura where the lobes contact one another and are lubricated with the same thin mucoid pleural fluid that lubricates the outer surface of the lungs. The lobes slide against each other in the same way that the entire lungs slide within the thoracic cavity. The diaphragm is the principle respiratory muscle. Contraction of the diaphragm elongates the lungs, forcing them to inflate. Other muscles elevate the anterior thoracic wall or compress the abdomen, raising the ribs from an inferiorly slanting position to a horizontal position that increases the anteroposterior diameter of the chest.

Returning to the airflow, bronchiole walls are composed of smooth muscle and connective tissue. This smooth muscle normally remains relaxed so that the bronchioles stay open, although particles of irritating substances entering these passageways can cause bronchiolar spasms (as in patients with asthma). The bronchioles branch several times. The last bronchiolar branch that still has a complete muscular coat is the terminal bronchiole. This too branches into several respiratory bronchioles with sparse smooth muscle. These bronchioles are the entryways into the respiratory lobules, the final air spaces of the lungs. The walls of all parts of the respiratory lobules form the respiratory membrane which totals ~70 m^2 in total surface area in both lungs combined. The membrane is so thin (<~1 micron) that oxygen and carbon dioxide can diffuse freely between the air inside the lobule and the blood in the capillary surrounding the lobule.

As in the nasal passages, the bronchial tree from the lower pharynx down to the end of the respiratory bronchioles is lined with ciliated columnar epithelial cells. These cilia also wave constantly toward the pharynx, moving secreted mucus at ~1.4 cm/min which replaces the entire mucoid coating once every ~20 minutes. A second protective mechanism is provided by mobile phagocytes present in the airways and in the alveoli, that engulf inhaled particles and bacteria and thus prevent this foreign matter from gaining access to other lung cells or from entering the blood by conveying this matter into the lymph system. Ciliary activity may be inhibited for several hours by smoking a single cigarette. Phagocytes are also injured by cigarette smoke, air pollution, and other noxious agents. Below the respiratory bronchioles, the ciliated columnar epithelium gives way to a nonciliated cuboidal epithelium.

Alveoli first begin to appear in the respiratory bronchioles, attached to the walls, and their frequency increases in the alveolar ducts until the airways end in grapelike clusters of alveoli. The alveoli are tiny hollow sacs 100-300 microns in diameter that open onto the lumina of the airways. Typically the air in two alveoli is separated by a single wall. The moist air-facing surfaces of the alveolar wall are lined by a continuous layer, one cell thick, of squamous type I epithelial cells. The alveolar surface contains smaller numbers of thicker specialized type II epithelial cells that secrete a detergent-like substance, or surfactant (a complex of protein with dipalmityl lecithin) in a fluid layer ~70 nm thick, which physically stabilizes alveoli of different sizes during inflation and deflation.

The alveolar walls also contain capillaries, the endothelial linings of which are separated from the alveolar epithelial lining only by a basement membrane and a very thin interstitial space containing interstitial fluid and a loose meshwork of connective tissue. In places, the interstitium is absent and the alveolar epithelium fuses with the capillary endothelium, with total thickness <~1 micron. In some of the alveolar walls there are pores that permit the flow of air between alveoli, an important route when the airway leading to an alveolus is occluded by disease.

At rest, ~4 liters/min of air enter and leave the alveoli, while 5.4 liters/min of blood, the entire cardiac output, flows through pulmonary capillaries which total ~2400 kilometres in length and ~150 cm^3 in volume. During heavy exercise, the air flow to the alveoli can increase to 120-160 liters/min and the blood flow to 25-30 liters/min. The volume of air taken in with each breath (resting tidal volume) is ~0.4 liter/breath at a respiration rate of 18/min (12-15/min during sleep). Volume flow may rise to ~3.2 liters/breath at a respiration rate of up to 62/min during the most strenous exercise. The two human lungs hold a combined ~6 liters of gas, of which ~3.7 liters is the maximum inspirational capacity leaving ~2.3 liters as residual capacity or dead space.

Blood pressure in the pulmonary artery is only 15-30 mmHg systolic and 4-12 mmHg diastolic, compared to 100-150 mmHg systolic and 60-100 mmHg diastolic in the aorta. This low-pressure system allows pulmonary capillaries to be quite flexible and thin-walled — they distend if higher blood pressures are applied to the pulmonary artery. Pulmonary blood vessels are richly supplied with nerve fibres but are largely free from neural and chemical control — although they do respond to hypoxia and to pharmacological doses of catecholamines, histamine and serotonin.

A navigational map of the airways can be surprisingly compact. Only $\log_2(300 \times 10^6) \sim 28$ bits are needed to uniquely name each alveolus, requiring 8.4×10^9 bits to store all 300 million addresses. A complete bifurcation map of the airways down to the last generation of respiratory bronchioles can be stored in just $\sim10^7$ bits. If numerous positions within each alveolus must be specified, during a scan seeking precancerous epithelial cells, the required map could be much larger. However, C. Phoenix notes that if nanorobots are given individual instructions to scan a certain set of cells for cancer, a contiguous tissue volume could be searched, specified by a small number of partial addresses. If 10^7 nanorobots were tasked to scan 30 (unnamed) alveoli each, within the contiguous volume, then each nanorobot must store one ~23 bit hard-coded kernel address plus one ~5 bit alveolus extension address (total 28 bits/alveolus) and one 30-bit cell address for each cancerous cell discovered, a modest memory requirement. On the other hand, general surveyor nanorobots lacking individualized instructions can keep track of the forks they pass and the branches they take as they enter the lungs, assembling the final address as they go. Upon reaching the terminus of the bronchial system and encountering another nanorobot already at work, the newcomer moves on to another location, keeping track of its current address (~28 bits). Once a cancer is detected, treatment nanorobots can follow the surveyor's address to return to the specific cancer cell without requiring a comprehensive map of the rest of the lung. Thus a treatment protocol can be executed without any overall map of the lung, although the assignment of individual search territories may prove more efficient, and multiple connectivities within the bifurcation tree may demand storage of additional bits or require post-survey data compression.

NAVIGATIONAL OSTEOGRAPHY

The skeleton is the single largest organ system, representing ~14 Percent of total mass and ~11 Percent of the navigable volume of the human body. The bony skeleton supports and protects the vital organs – the skull protects the brain; the spinal column shields the spinal cord and maintains erect posture; the ribs shelter the heart, lungs, and liver; pelvic bones protect the kidneys and internal sexual organs.

The total number of bones varies at different ages. At birth, the human body contains ~270 bones. This number declines slightly during infancy as a few separate segments join to form single bones. From young childhood through puberty, the bone count increases as wrist and ankle bones develop. Post-adolescence, the bone count steadily declines again with the gradual union of independent bones.

The adult skeleton consists of 206 bones: 28 skull bones (8 cranial, 14 facial, and 6 ear ossicles); the horseshoe-shaped hyoid bone of the neck; 26 vertebrae (7 cervical or neck, 12 thorax, 5 lumbar or loins, the sacrum which is five fused vertebrae, and the coccyx, our vestigial tail, which is four fused vertebrae); 24 ribs plus the sternum or breastbone; the shoulder girdle (2 clavicles, the most frequently fractured bone in the body, and 2 scapulae); the pelvic girdle (2 fused bones); and 30 bones in each of the four extremities (a total of 120). The paired bones include the 12 ribs on either side, 8 wrist bones, 5 hand bones, 14 finger bones, 7 ankle bones, 5 foot bones, 14 toe bones, 3 auditory ossicles, and the parietal, temporal, palatine, lacrimal, nasal, upper jaw, cheek, lower nasal concha, collarbone, shoulder blade, hip, arm, outer and inner forearm, thigh, kneecap, calf, and shin bones. There are also a variable number of sesamoid bones, ranging from 8-18 in number, which are small rounded masses embedded in certain tendons and usually related to joints.

The total mass of the skeleton is ~10,000 gm; skeletal density averages ~1.45 gm/cm^3, so total volume is ~6875 cm^3. The total exterior surface area of the bones is ~1 m^2, which may be mapped to ~(20 micron)^2 cellular resolution using 2.5 billion pixels, requiring ~0.02 terabits of onboard nanorobot memory assuming 8-bit pixels.

More interesting are the interior spaces of bones in which medical nanodevices may travel. Bone is composed mainly of ~50 nm long crystals of very dense calcium salts having the hardness of marble (mostly hydroxyapatite, ~45 Percent of bone mass) held together by bone matrix containing large numbers of extremely strong ~200 nm-wide collagenous fibres arranged in a dense mat together with a mucopolysaccharide cement sometimes called "ground substance." In living bone, ~25 Percent of bone weight is water and another ~30 Percent is organic material.

Embedded in this mineral structure are bone cells of three types. Osteoblasts, which secrete the substances that make up the bone matrix, line the outer surfaces of bone and also line many of the surfaces inside the internal cavities of the bone. Osteocytes, the most numerous bone cell type, originate from osteoblasts when those cells become trapped within small irregular matrix cavities called lacunae. After the trapped cells transform into osteocytes, they stop forming new bone but continue to support normal bone metabolism. Finally the osteoclasts (large multinucleated cells lining ~3 Percent of internal bone cavity surfaces) remove old bone when it needs repair. Osteoclasts may be stimulated by parathyroid hormone to cause bone absorption when extra calcium ions are needed in the extracellular fluid. Bone is continually being resorbed and rebuilt to maintain its structural integrity.

Compact bone is formed on the exterior of all bones. Deeper inside, the bone becomes a more open latticework with large spaces, called spongy or cancellous bone, whose solid parts nevertheless have an internal structure similar to compact bone. In the long bones of the extremities, the epiphyses are composed of spongy bone covered by a thin layer of compact bone. The diaphysis is made up almost entirely of compact bone surrounding a cavity containing marrow. Covering the entire bone is the periosteum, a nutrient-carrying fibrous membrane investing the surfaces of bones, except at the points of attachment of tendons and ligaments. The periosteum has an outer fibrous layer composed of dense fibrous tissue with blood vessels and an inner osteogenic layer containing many fibroblasts. Blood vessels traversing the periosteum enter the bone and pass through channels called Volkmann's canals to enter and leave the Haversian canals. In humans, there are few true voids in the bones.

Bone marrow fills the spaces of spongy bone and the medullary cavity of long bones. It is composed of a supporting framework of reticular tissue in which there are blood vessels and blood cells in various stages of development. A thin membrane called the endosteum, more delicate than the periosteum but resembling it in structure, lines the medullary cavity. In the adult, there is red and yellow marrow. Red cells and some white cells are formed in the red marrow; in the newborn all marrow is red, but in the adult the red marrow is found in the spongy bone such as the proximal epiphyses of long bones and in the sternum, ribs, vertebrae and diploe of the cranial bones. Yellow marrow contains many fat cells and is found in the medullary cavity of long bones, producing macrophages and granulated white cells. There is ~3 kg of marrow in the adult body, occupying a volume of ~2400 cm^3.

Even the relatively low capillary density of <100/mm^2 in bone implies a total investiture of ~3×10^8 skeletal capillaries and a total luminal surface area of ~8 m^2 for the entire vasculature, which would require a ~0.16 terabit vascular map at cellular resolution. A full cellular map identifying each individual osteocyte in the entire skeletal system requires ~1 terabit.

An interesting navigable volume within the skeletal system is the human spine, which averages 71 cm in length in men and 60 cm in women with remarkably little variation. The body of each vertebra has a ring-shaped neural arch, through which passes the spinal cord. The spinal cord, measuring ~46 cm long and ~1 cm in diameter, descends from the brain and medulla oblongata through the foramen magnum and passes through the hollows of the spinal column, giving off 31 pairs of spinal nerves until it reaches the level of the disc between the first and second lumbar vertebrae, where the lower end tapers off to a point called the conus medullaris. The cord itself is composed of H-shaped gray matter in the centre, extending forward as the anterior nerve roots and backward as the posterior nerve roots, with solid white tracts of nerve fibres descending from the brain or ascending to it.

Cerebrospinal fluid fills the ventricles of the brain and the subarachnoid cavity surrounding the spinal cord. The fluid is formed mainly by the choroid plexuses in the four ventricles of the brain. The plexuses are rich in blood vessels and separated from the cavity of the ventricles by only a single layer of secretory cells, thus affording ready access to the fluid by medical nanorobots. The fluid, once formed in the brain, is later mostly reabsorbed by the arachnoid villi in the brain, but a small amount drifts slowly downstream along the spinal cord and is absorbed by the arachnoid villi through the spinal regions. Among other things, CSF is regarded as "the drainage system of the brain," crudely analogous to urine, and the entire fluid volume is replaced once every ~30,000 sec.

Most of the ~150 bone joints in the human body are freely movable diarthroses. There are ball and socket joints, saddle joints, ~40 hinge joints, and pivot joints. In the diarthrodial joint, two or more bones are united by an encircling band of fibrous tissue called the articular or fibrous capsule. The articular capsule is lined with synovial membrane, and the apposed ends of bone are covered by a layer of hyaline cartilage, called articular cartilage. The fibrous capsule is reinforced and strengthened by ligament cords woven into it. Joint cavities sometimes extend into pouches or recesses or communicating bursae, and the shape of the sac changes with motion, as in the knee joint.

There are also ~140 bursae in the body, located in fibrous tissue wherever friction occurs, probably containing at least ~20 cm^3 of additional synovial fluid. The best-known include the subdeltoid bursa, which lies beneath the shoulder muscle; the prepatellar bursa, located in front of the kneecap; and the Achilles bursa, which lies between the heel bone and the Achilles tendon at the back of the heel.

Positional Navigation

Positional navigation should allow a medical nanorobot to know its three-dimensional coordinate position to ~micron accuracy at all times. A nanodevice may rely upon dead reckoning, cartotaxis, microtransponder network alignment, or triangulation on external beacon signals.

Cartotaxis

Another positional navigation technique is simple landmark-centered map-following, or cartotaxis. Consider a nanorobot that wishes to crawl to a specific 1 $micron^2$ patch on the subvilliary mucosal surface of the small intestine. Ignoring the villi and microvilli but including the circular folds, the mucosal surface of the small intestine is ~1 m^2, at worst requiring a map of ~8 terabits to achieve 1 $micron^2$ resolution using 8-bit pixels.

A more compact map might include only the positions, sizes and shapes each of the ~100 million 50-micron-wide intestinal glands dotting the surface of the small intestine at least ~100 microns apart. These dots form a pattern unique to each person that changes slowly over time. After defining a two-dimensional intestinocentric coordinate system, two coordinates of $\log_2 (10^6) \sim 20$ bits each can specify a location to 1-micron accuracy on a 1 m^2 surface. Using 20 bits each for the two surface coordinates, 20 bits each for the major and minor elliptical axes framing the gland aperture, and another 920 bits to record unique topographic features of each hole, a complete small intestinal gland map requires at most ~1000 bits/gland or ~0.1 terabits.

This data storage requirement may be significantly further reduced by recognizing that on its way to its destination, an efficient ~10 $micron^3$ surface-walking nanorobot need pass over at most 0.002 Percent of the 1 m^2 of small intestinal surface recorded on the map described in the previous paragraph, even assuming a maximum-length 4-meter whole-intestine traverse, passing at most a total of 4 m / 100 microns=40,000 identified glands each requiring ~1000 bits to describe. Hence the minimum intestinoglandular map required for this longest traverse could in theory contain as few as ~40 megabits of data for a perfectly navigating nanorobot using a perfectly compiled and absolutely stable map. As a practical matter, to ensure reliability the onboard map conservatively should contain complete intermediate annular ring segment recalibration maplets at least ~500 microns in length along the tube axis and spaced ~2 cm apart, to be sequentially encountered axially while traveling down the tube. Nanorobots use dead reckoning to navigate between these guide rings until the ring closest to the destination is reached, analogous to the mid-course correction areas employed by cruise missiles. Having arrived at this nearest ring, the nanorobot follows a final map swath ~500 microns wide leading directly to the three glands lying closest to the destination patch; the final 1 micron^2 target patch is reached by interpolation and dead reckoning between these last three glands. Reliability and efficiency are enhanced by using large numbers of intercommunicating nanorobots simultaneously. This ~1 gigabit "practical map" contains complete descriptions of ~1 million individual intestinal glands and may be stored in a ~0.1 micron^3 onboard data spool using hydrofluorocarbon memory tape.

The above scenario describes the ideal case. In actual living systems, biological surfaces are constantly in motion, are coated with slime, and are frequently being remodeled. Some mucosal surfaces may replace their entire luminal cell population every ~10^5 sec (~1 day). Thus the precise shapes and features of individual intestinal glands are constantly changing, although their larger structures and the pore pattern may remain intact for considerably longer periods of time. Intestinal glands are easily located by moving in the direction of rising concentration of enzyme-loaded intestinal juice, a basic chemonavigational technique, and villiary trunks are available to provide additional cartographic guidance. But cartotaxic nanorobots will achieve the best results by using only the most recently prepared maps.

Dedicated Navigational Organs

By direct analogy to communication organs, dedicated macroscopic organs may be implanted in the human body to facilitate navigation. One important function of such organs would be to act as central clearinghouses for systemic information that is normally available only regionally or to individual navicytes. This information might include updated positions of all regional monuments, orientations of various grid sectors relative to gravity, or information on macroscopic rotation rates to allow centrifugal and gravitational forces to be distinguished. Highly accurate chronometers could also be maintained in these organs to allow periodic systemwide resynchronization.

Dedicated navigation organs could serve as the "map rooms" of the body, collating and organizing new information as it comes in, maintaining accurate navicyte grid maps, organ maps, vascular maps, functional maps, and so forth, possibly in coordination with dedicated computational organs. A 1 mm^3 navigational library node can contain up to ~10 million terabits of map data, which may be downloaded to nanorobots berthed at docking ports at rates of up to 0.01 terabit/sec drawing ~50 pW per docking port during the transfer.

Dedicated navigation organs may be employed as direct interfaces between internal navigational systems and related macroscopic external modalities including operating theaters, clinical equipment, various environmental entities, satellite uplinks, radio antennae, transportation vehicles, and the like. They could also serve as emitter organs in GPS-like acoustic navigational systems.

FUNCTIONAL NAVIGATION

Functional navigation allows a medical nanorobot to detect and respond to subtle variations in tissue characteristics, often in the absence of precise positional knowledge. Such tissue characteristics may include thermal, acoustic or barostatic, cytochemical or immunochemical, electrical or magnetic, mechanical or topological variations.

Thermographic Navigation

The human body presents a complex and temporally varying spatial temperature field. The external parts of the body have a lower mean temperature than the internal parts, with temperature decreasing along the longitudinal axis of the extremities, producing both axial and radial temperature gradients. The differing heat production of individual organs, geometric irregularities, changes in insulation and evapouration, convective heat transport via the blood, and the diurnal and other periodic variations add further complexities to the thermal map.

The homeothermic core of the body is distinguishable from the shell, which most readily responds to environmental fluctuations. The core generally consists of the interior of the thorax and abdomen, the brain, and part of the skeletal muscles. With moderate changes in ambient temperature, the shell normally comprises the outermost 20 Percent-35 Percent of the human body. However, during extreme chilling, the shell may enlarge to ~50 Percent of total body volume, equivalent to a mean layer thickness of 2.5 cm.

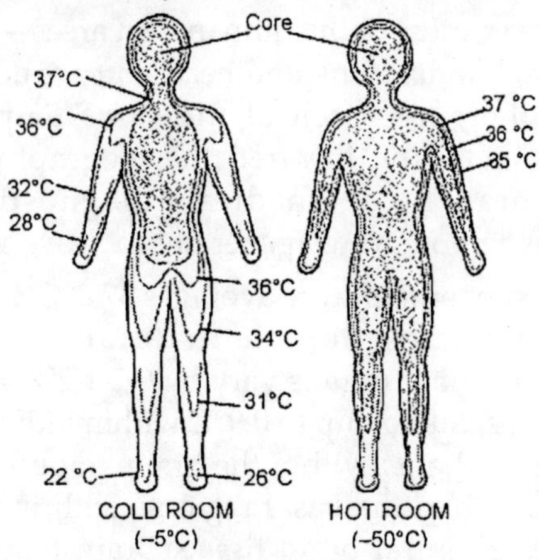

Fig. Distribution of isotherms in human body placed in cold or hot environment

Skin temperatures display the greatest thermographic variability in response to external factors. Nude humans standing for 3 hours in a cold room (5°C, 50 Percent relative humidity, 0.1-0.2 m/sec wind speed) experience skin temperature differentials up to 15°C (=13°C to 28°C), with the lowest temperatures in fingers and toes, the highest in trunk and forehead, and average core/surface gradient ~15°C; heat loss is ~10 Percent lower for females due to their thicker layer of subcutaneous fat, making their cold-room skin temperature slightly lower than for males. After 3 hours in a hot room (50°C), skin temperature differentials amounted to only 2.5°C (=35°C to 37.5°C), with an average core/surface gradient of ~1°C. With normal clothing in a room at 15-20°C, mean skin temperature is 32-35°C.

Skin thermography of the human head was first reported by Edwards and Burton; skin vs. rectal temperatures at various ambient temperatures are well-studied. Skin temperature patterns in neonates reflect near uniform heat conduction through the tissues. In childhood, specific patterns develop into a stable, permanent adult pattern. The dermothermal patterns differ in lean and obese patients, and exhibit a continual state of small rhythmic change. These changes — probably a result of active vasodilation due to sympathetic innervation over most of the human skin area — are observed over the arms, hands, trunk and head, but are not all in phase with each other, nor even of the same amplitude. Skin thermology, including infrared thermography, is now an important branch of medical diagnostic imaging. Subcutaneous temperatures generally increase with depth.

Human core temperature averages 37.0°C, but this simple number hides considerable natural variation. The temperatures of inner organs vary by 0.2-1.2°C under normal room conditions, and by up to 0.9°C within individual organs. Temperature gradients within the brain amount to 1.4°C; the cortex is cooler than the basal regions, with incoming blood cooler than the central brain tissue. Brain temperature also decreases during sleep and rises during periods of emotional arousal. Cooling or warming the skin of the head causes temperature changes in the tympanic membrane of up to 0.4°C due to the returning venous blood. Oral temperature can have wide variation, depending on the thermal character of recently ingested food and drink, and thus has little direct correlation with swings in rectal temperature. Liver temperatures lie 0.2-0.6°C subrectal, possibly due to heat loss via the respiratory tract through the diaphragm. The temperature of the testes in the scrotum is ~3°C lower than that of the abdominal cavity. The temperature of the hypothalamus, the seat of human thermoregulation, might be the most logical choice as a reference value.

Human blood temperature also displays considerable variation. Outflowing blood is warmer than inflowing blood in organs with high metabolic rates, so the blood has a cooling effect in these organs, the opposite of the skin. Venous blood temperature in the extremities may be ~3°C lower than the arterial supply, but even the arterial blood can sink to 22°C in human extremities at cold room temperatures. Blood in the common carotid artery remains within 0.2°C of oral temperature, but falls 0.2-0.5°C as it cools the face and possibly also due to cooling from the internal jugular vein. However, inhalation of cold air has little effect on pulmonary blood temperature; even for patients in rooms cooled to 18°C, blood in the pulmonary vein and artery differ at most by 0.03°C.

There are also periodic variations in body temperature. The diurnal fluctuation is the best known. This cycle is absent in neonates, but develops during the first weeks of life. Children have higher daytime temperatures than adults and a daily range of 1.7°C, with the daily range peaking near the age of 1 year. Precisely standardized tests on young adults found that rectal core temperature fluctuated between 36.83-38.32°C in males and 37.16-38.36°C in females, with temperatures normally reaching their lows near 6 AM local time and their highs near 6 PM. The menstrual cycle also involves a complex long period fluctuation. During this monthly cycle, morning core temperature falls from 37.0°C to 36.7°C just before and during menstruation, after which there is an abrupt dip to ~36°C at mid-period near ovulation, followed by a return to normal levels for the remaining ~2 weeks. Menstrual-related skin temperature pattern variations are also observed.

Finally, there are various irregular influences on body temperature. Rectal temperature may rise up to 3.5°C during the most extreme exercise, or may fall 2°C after ~1 hour immersion in water at 23°C. Ingesting food elevates skin temperature, and to a lesser extent the rectal temperature, for 1-2 hours. Alcohol ingestion raises skin temperature but lowers core temperature, while nicotine ingestion lowers skin temperature in the extremities by 2-7°C. In persons with fever, the temperature may rise from 37.0°C up to 41.5°C, with daily fluctuations of 2.8-5.0°C, usually peaking in late afternoon. Transient elevations to 44-45°C have been recorded but are very rare.

In view of this measurable spatial and temporal variability, maintaining a high-resolution whole-body thermal map in real-time is impractical using in vivo nanodevices. Given the availability of ~microkelvin sensors with ~microsecond measurement cycles, sensor technology is not the limiting factor. Map stability is the main concern. A volumetric map with only ~1 mm^3 resolution, comparable to the best thermographic images, requires 10^8 voxels and 17 bits/voxel for 0.001°C accuracy. A hand placed on a stove produces thermal gradients >10°C/sec in the shell, but periodic temporal variations are 0.1-40 microkelvins/sec in the core and 0.0051°C/sec in the shell during changes in physical exercise levels, movement between hot and cold environments, and fevers. Thus to maintain a map that reports all changes of >~1 millikelvin requires the entire field to be resampled every 0.001°C / (1°C/sec)=1 millisec, generating ~2 terabits/sec. This dataflow possibly could be accommodated by a dedicated high-capacity fibre network but would rapidly overwhelm both a mobile acoustic network and the data storage capacity of individual bloodborne nanorobots. Maintaining a hematothermographic map employing capillary-volume resolution to millikelvin accuracy generates an even more intractible >10^{15} bits/sec.

Thermal Demarcation

Thermographic navigation to a target site is possible without the use of any maps. Simple thermal demarcation is a good example of this. In a controlled room-temperature clinical setting, the relatively narrow homeostatically-maintained temperature band of 36-38°C allows easy demarcation of selected target regions by applying thermal stimuli lying just outside the normal band. Examples might include hot (T > 38°C) or cold (T < 36°C) packs applied to the skin; diathermic tissue heating using different frequencies to achieve selective heating at various tissue depths; adaptive phased array heating of small target volumes deep inside the body; thermally active catheter probe tips inserted transdermally or otherwise, directed to specified internal targets; or multiple intersecting focused infrared, visible photon, or ultrasound beams directed to specific convergence volumes for maximum heating effect. Depending upon the technique employed and the target tissue, volumes as small as ~1 mm^3 may be demarcated in this manner.. Nanorobots of relatively simple design may be injected near the target or allowed to circulate in the bloodstream, and will detect the outlier temperature data and gather in the vicinity of the target at a statistically determined accumulation rate and number density.

Outside of controlled environments or during prolonged treatment periods, simple demarcation becomes less reliable for several reasons:

1. Error in the physical placement of the instrument of demarcation will produce erroneous nanorobot localization, which may increase the likelihood of treatment error. Such iatrogenic mistakes are especially likely during time-critical or emergency crises, or in situations where the diagnosis is unclear, or in cases of self-administered medical care.
2. Even when initially correctly placed, the artificial hot or cold region will diffuse outward from the original spot, enlarging the radius of action in imprecisely known directions, again increasing the risk of iatrogenic effects.
3. Load error in the human thermoregulatory control system is ~0.2-0.5°C. Any thermal deviation of this magnitude or larger from local temperatures will provoke a natural counteractive response to restore the local temperature, including capillary sphincter action, increased sweat gland secretions, accelerated counter-current exchanges, more rapid heart rate, or local thermogenesis.
4. Increased susceptibility to spoofing, as when a patient enters a hot shower or bath, lies down on snow or on a cold concrete surface, washes his hands and face with hot water, drinks a hot or cold beverage, basks in the sun, and so forth. Temperatures hot or cold

enough to definitively prevent spoofing might damage tissues if maintained for very long. This problem can be avoided by employing an oscillating artificial thermal gradient instead of a time-invariant source, but at the cost of increased required overall nanorobot sophistication to detect the oscillating demarcation. For energy delivered to a target L ~ 1 mm from the source in a medium of heat capacity C_V and thermal conductivity K_t, the maximum frequency that may be resolved is $n_{thermal} \sim K_t / C_V L^2 \sim 0.15$ Hz for aqueous tissue at 37°C.

General positional information is also available to nanorobots equipped with thermal sensors in the absence of any maps. Assuming the temperature of the external environment has been measured and disseminated via the communication network, a nanorobot traveling within a microvolume of blood can crudely infer its distance from the epidermis based on the local blood temperature, especially when passing through the thermal shell of the body surrounding the core. Other cues including velocity measurements or histonavigational data may allow further refinement of the estimate.

BAROGRAPHIC NAVIGATION

The internal pressure field of the human body is as complex as the temperature field. In general, fluid pressures range from 60-150 mmHg in the heart, arteries and arterioles; ~30 mmHg in the capillaries and 5-20 mmHg in the veins; 4-30 mmHg in the pulmonary artery; 9-13 mmHg in the cerebrospinal and synovial fluids; ~1.4 mmHg in the interstitial fluid between cells; and ~0.9 mmHg mean pressure in the lymphatics, with arterial pulsations contributing 1.5-2.2 mmHg oscillations with a maximum of ~5 mmHg in the lymphatics.

As in thermographics, simple barographic demarcation provides a ready means for directing medical nanorobots to a desired treatment location within the body. Simply pressing down on the surface of the skin creates an anomalous region of elevated tissue pressure; tapping, slapping, or palpitating the body's surface is also detectable by in vivo nanorobots. Acoustically active catheters inserted transdermally can place a sonic beacon signal near almost any target location. Highly focused but relatively low power ultrasonic columnar or planar beams of, say, three different frequencies can be directed into the body on various trajectories and from various sources. The tiny spot where all three frequencies are audible at precisely equal intensity marks the target, providing <~1 mm^3 localization accuracy at frequencies >~1.5 MHz. Focused ultrasound surgery (FUS) and high-intensity focused ultrasound (HIFU) may use ~2-second bursts of 1.7-MHz, 150-mm-deep focused beams at 5×10^7 watts/m^2 to occlude blood flow or destroy tumors by creating a spot of intense heat so small that there is a boundary of only ~6 cells between destroyed tissues and completely unharmed tissues, a ~100 micron spatial resolution.

As in the case of temperature fields, low-resolution and high-resolution barographic maps can be compiled and stored by barographicytes, then used to diagnose a wide variety of medically relevant states in organs and throughout the body including tumors, edemas, excess cranial pressure, general hypertension or hypotension, swollen lymph glands, muscle spasms and the like. Such maps are also useful in macrosensing, as for instance to help determine if the patient is sitting, standing, prone, inverted, falling, or floating. Blood pressure increases in patients who are exercising, eating a meal, or emotive, and decreases in patients who are relaxing or sleeping. The typical daily readings in a hypertensive 35-year-old male might be 150-220 mmHg systolic and 90-140 mmHg diastolic.

However, the amplitude and timing of pressure waves in the blood vessels permit somewhat clearer positional inferences than is possible with thermographics, in part because the heart is such a strong, reliable, and stably-positioned acoustic pulse generator.* Functional hemobaric mapping thus may provide useful positional information regarding vascular range estimates.

CHEMOGRAPHIC NAVIGATION

The human body is a cauldron of chemical complexity. Of particular interest to us here are those extracellularly occurring chemical species reliably associated with specific locations, functions, or processes that might be useful in navigation. As trivial examples, high concentrations of hormones or neuropeptides are found in the endocrine glands or in the nerve tissues; lactic acid is produced in muscles driven anaerobically; uric acid is heavily concentrated in urine in the bladder; capillary lymph is more watery than thoracic lymph, which in turn is far richer in histaminase than cervical duct lymph.

As with thermal and pressure markers, crude chemical demarcation markers may be employed to direct medical nanorobots to particular locations. Such markers may include injected chemical plumes, time release implants, remote-triggered releasers, and tissue-coded markers — all of which may establish useful localized chemical gradients. The principle difficulty for natural signal molecules is signal persistence, which may be relatively brief unless continuous broadcast and artificial signal molecules are employed. Continuous broadcast creates a detection sphere of maximum radius R_{max} around the broadcast emission point. Use of artificial non-biodegradable molecules maximizes the signal/noise ratio. Diffusion causes small molecules to drift ~1 cell width in ~1 second, or ~1 mm in ~1000 seconds; bulk flow microconvection and increased effective diffusivity due to natural stirring further limit positional precision.

Use of chemical beacons for chemonavigational triangulation is not efficient over large distances, since the required volume of signal chemical rises as the cube of the range. However, chemical localization may be moderately useful in close quarters. A micron-size nanorobot equipped with two antipodal chemical sensors can angularly resolve two periodic sources of dissimilar signal molecules to the limit of nanodevice rotation during a Brownian motion-limited measurement time. A pair of 10-probe sensors taking measurements every ~1 millisec can resolve Da ~ 3° of arc, or two chemical point-emitters separated by X_{path} sin(Da) ~ 1 micron at a range of X_{path} ~ 20 microns (~1 cell width) from the nanorobot.

Chemonavigation can also provide useful positional information. Glucose absorption in the jejunum establishes chemical gradients both in the radial and axial directions in the lumen and also across the surface of the small intestine. If chyme leaving the duodenum starts with 1 Percent glucose concentration and exits the upper jejunum at 0.01 Percent concentration ~50 cm downstream, then the same 10-probe chemical sensor as above can measure 25 distinguishable glucose levels as a function of distance from the pylorus. This allows jejunal position to be estimated to within ~2 cm using glucose concentration alone.

Chemical navigation in the gut may incur significant positional uncertainties due to the large chemical variation of ingested foodstuffs, the generation and transport of gases, the volume and character of liquids consumed with the food, the state of mind and bowel peristalsis of the patient, the presence of biochemistry-altering disorders, and other factors. On the other hand, the results of measurements of nutrient levels in the stomach and duodenum may be broadcast downstream to jejunal nanorobots, allowing real-time assessment of the efficiency of digestive processes and thus reducing positional uncertainties. Similar gradients allow positional chemonavigation in gland ducts such as the salivary and pancreatic ducts.

Sulfate groups are more numerous on the surfaces of the rabbit aortic arch and carotid than on the surfaces of the rest of the systemic arteries, and post-staining fluorescence intensity measuring abundance of sialyl groups in the endothelial glycocalyx is 1.65 times higher in the carotid than in the aorta. In cyto chemonavigation is possible as well. Intracellular diffusion gradients of O_2 and ATP that appear under some conditions could be useful for detecting the distribution and clustering of mitochondria.

Detailed chemomapping of individual patients may allow customization of standardized human chemographic profiles, e.g., to identify problem areas. Chemographicytes may assist in the mapping process at up to ~KHz sampling frequencies. Regions of hypoxia develop in all solid tumors where rapidly dividing cells are supplied by inadequate or poorly developed vasculature. A continuously updated tissue oxygenation map allows recognition of growing hypoxic regions, permitting prompt detection of nascent tumors, myocardial and cerebral ischemia, cardiorespiratory failure, and rheumatoid arthritis. Similarly, a whole-body map of the concentration of intracellularly sampled telomerase or associated proteins such as TRF1/TRF2, Ku, and tankyrase, or even telomerase mRNA levels may provide a good first-cut toward a whole-body late-stage cancer map, since 85 Percent-90 Percent of all primary tumors display this biochemical marker. Cancer cells also display above-normal concentrations of b_1 integrins, survivin, sialidase-sensitive cancer mucins and leptin receptors such as galectin-3, and below-normal concentrations of b_4 integrins. Other examples include GM2 ganglioside, a glycolipid present on the surface of ~95 Percent of melanoma cells, with the carbohydrate portion of the molecule conveniently jutting out on the extracellular side of the melanoma cell membrane. GM2 and another ganglioside, GD2, are expressed in several types of cancer cells including small-cell lung, colon, and gastric cancer, sarcoma, lymphoma, and neuroblastoma. Numerous specific markers of disease are available for detection and mapping by nanorobots.

MICROBIOTAGRAPHICS

While high-speed nanomedical prophylactic measures against pathogens will be readily available, in some cases the physician may wish to obtain an infection map before beginning treatment. It may also be useful to prepare maps of particular symbiotic, commensal, or parasitic bacterial species, or native mobile cellular elements, in the human body, whether for scientific or for diagnostic purposes.

Leukocytes and macrophages are readily recognized and could be statistically sampled to produce a crude whole-body leukographic map. A fleet of 2 trillion mapping nanorobots evenly distributed throughout a 0.1 m^3 body volume gives each nanorobot a patrol volume of 50,000 $micron^3$. At 310 K, a 1-micron nanorobot of normal density has an average thermal velocity of ~500 microns/sec. Assuming a cross-sectional area of ~1 $micron^2$, in 1 second a motile device traveling at this speed may trace out a nonrepeating zigzag path ~500 microns in length, or ~1 Percent of its patrol volume, even while diffusing an additional radial distance of only DX ~ 1 micron. During its zigzag course, the device may collide at least once with ~1 Percent of the motile body cells or microbiota present within the patrol volume. If ~10 collisions are required to ensure a positive identification of a cellular coat, then ~0.1 Percent of the entire target cell population present in the patient's body may be physically sampled, positively identified, affirmatively counted, and directly associated with a specific map voxel in ~1 sec – a sufficient statistical sample for most purposes, even given the likelihood of substantial double-counting. A whole-body white-cell-count map to ~1 cm^3 voxel resolution requires ~10^6 bits to describe and may be retrieved by the physician in <~1 sec using an in vivo mobile acoustic communication network; a ~1 mm^3 resolution map (~10^9 bits) fine enough to discover even the smallest macroscale infection site requires ~100-1000 seconds to outmessage in this manner. A sequence of readouts at regular intervals provides a clear strategic overview of the developing infection – some bacterial infections cause a selective increase in neutrophils, while infections with some protozoa and other parasites cause a selective increase in eosinophils.

Whole-body vascular surveys of red cells, white cells, platelets or pathogens will proceed ~10 times faster using a nanorobot fleet of similar number density, owing to the reduced search volume, though at the cost of slightly increased onboard computational complexity needed to compensate for the moving reference frame.

IDENTIFICATION OF CELL TYPE

The usual estimate based on histological studies is that there are ~200 distinct kinds of cells that show alternate structures and functions. These represent discrete categories of cell types of markedly different character, not arbitrary subdivisions along a morphological continuum. Traditional classification is based on microscopic shape and structure, and on crude chemical nature, but newer immunological techniques have revealed, for instance, that there are more than 10 distinct types of lymphocytes. Pharmacological and physiological tests have revealed many different varieties of smooth muscle cells — uterine wall smooth muscle cells are highly sensitive to estrogen and oxytocin, while gut wall smooth muscle cells are not. The catalog is organized by cellular function and omits subdivisions of smooth muscle cells, neuron classes in the CNS, various related connective tissue and fibroblast types, and intermediate stages of maturing cells such as keratinocytes only the stem cell and differentiated cell types are given. Otherwise, the catalog is said to represent an exhaustive listing of the ~219 cell varieties found in the adult human phenotype.

Medical nanorobots can probably distinguish all of these cell types by surface chemical assay using chemosensor pads. Antigenic specificities exist for species, organ, tissue or cell type for almost all cells — possibly involving as many as $\sim 10^4$ distinct antigens. A comprehensive analysis of the cytoimmunography of all cell types is beyond the scope, but a few examples of cell type-specific antigenic markers can be given to illustrate the tremendous power of this approach.

In the case of red blood cells, antigens in the Rh, Kell, Duffy, and Kidd blood group systems are found exclusively on the plasma membranes of erythrocytes and have not been detected on platelets, lymphocytes, granulocytes, in plasma, or in other body secretions such as saliva, milk, or amniotic fluid. Thus detection of any member of this four-antigen set establishes a unique marker for red cell identification. MNSs and Lutheran antigens are also limited to erythrocytes with two exceptions: GPA glycoprotein also found on renal capillary endothelium, and Lu^b-like glycoprotein which appears on kidney endothelial cells and liver hepatocytes. In contrast, ABH antigens are found on many non-RBC tissue cells such as kidney and salivary glands. In young embryos ABH can be found on all endothelial and epithelial cells except those of the central nervous system. ABH, Lewis, I and P blood group antigens are found on platelets and lymphocytes, at least in part due to adsorption from the plasma onto the cell membrane. Granulocytes have I antigen but no ABH.

Platelets also express platelet-specific alloantigens on their plasma membranes, in addition to the HLA antigens they already share with body tissue cells. Currently there are five recognized human platelet alloantigen systems that have been defined at the molecular level. The phenotype frequencies given are for the Caucasian population; frequencies in African and Asian populations may vary substantially. For instance, HPA-1b is expressed on the platelets of 28 Percent of Caucasians but only 4 Percent of the Japanese population. A medical nanorobot that detects a representative antigen from any of these five HPA antigen groups on an encountered surface can be certain that it has found a platelet.

Lymphocytes with a particular functional activity can be distinguished by various differentiation markers displayed on their cell surfaces. All mature T cells express a set of polypeptide chains called the CD3 complex. Helper T cells also express the CD4 glycoprotein, whereas cytotoxic and suppressor T cells express a marker called CD8. Thus a nanorobot that detects the phenotype $CD3^+CD4^+CD8^-$ has positively identified a helper T cell, whereas the detection of $CD3^+CD4^-CD8^+$ uniquely identifies a cytotoxic or suppressor T cell. All B lymphocytes express immunoglobulins (their antigen receptors, or Ig) on their surface and can be distinguished from T cells on that basis, e.g., as Ig^+ MHC Class II^+.

Lymphocyte surfaces also display distinct markers representing specific gene products that are expressed only at characteristic stages of cell differentiation. Stage I Progenitor B cells display $CD34^{+}PhiL^{-}CD19^{-}$; Stage II, $CD34^{+}PhiL^{+}CD19^{-}$; Stage III, $CD34^{+}PhiL^{+}CD19^{+}$; and finally $CD34^{-}PhiL^{+}CD19^{+}$ at the Precursor B stage.There are neutrophil-specific antigens and various receptor-specific immunoglobulin binding specificities for leukocytes. Monocyte FcRI receptors display the measured binding specificity $IgG1^{+++}IgG2^{-}IgG3^{+++}$ $IgG4^{+}$, monocyte FcRIII receptors have $IgG1^{++}IgG2^{-}IgG3^{++}IgG4^{-}$, and FcRII receptors on neutrophils and eosinophils show $IgG1^{+++}IgG2^{+}IgG3^{+++}$ $IgG4^{+}$. Neutrophils also have b-glucan receptors on their surfaces.

Tissue cells display specific sets of distinguishing markers on their surfaces as well. Thyroid microsomal-microvillous antigen is unique to the thyroid gland. Glial fibrillary acidic protein is an immunocytochemical marker of astrocytes, and syntaxin 1A and 1B are phosphoproteins found only in the plasma membrane of neuronal cells. Alpha-fodrin is an organ-specific autoantigenic marker of salivary gland cells. Fertilin, a member of the ADAM family, is found on the plasma membrane of mammalian sperm cells. Hepatocytes display the phenotypic markers $ALB^{+++}GGT^{-}CK19^{-}$ along with connexin 32, transferrin, and major urinary protein, while biliary cells display the markers $AFP^{-}GGT^{+++}CK19^{+++}$ plus BD.1 antigen, alkaline phosphatase, and DPP4. A family of 100-kilodalton plasma membrane guanosine triphosphatases implicated in clathrin-coated vesicle transport include dynamin I (expressed exclusively in neurons), dynamin II (found in all tissues), and dynamin III (restricted to the testes, brain, and lungs), each with at least four distinct isoforms; dynamin II also exhibits intracellular localization in the trans-Golgi network.

Bacterial membranes are also quite distinctive, including such obvious markers as the family of outer-membrane trimeric channel proteins called porins in Gram-negative bacteria like and other surface proteins such as *Staphylococcal* protein A or endotoxin, a variable-size carbohydrate chain that is the major antigen of the outer membrane of Gram-negative bacteria. Mycobacteria contain mycolic acid in their cell walls.

In addition, only bacteria employ right-handed amino acids in their cellular coats, which helps them resist attack by digestive enzymes in the stomach and other organisms. Peptidoglycans, the main structural component of bacterial walls, are cross-linked with peptide bridges that contain several unusual nonprotein amino acids and D-enantiomeric forms of Ala, Glu, and Asp. D-alanine is the most abundant D-amino acid found in most peptidoglycans and the only one that is universally incorporated.

At least four major families of cell-specific cell adhesion molecules had been identified by 1998 – the immunoglobulin (Ig) superfamily (including N-CAM and ICAM-1), the integrin superfamily, the cadherin family and the selectin family.

Integrins are ~200 kilodalton cell surface adhesion receptors expressed on a wide variety of cells, with most cells expressing several integrins. Most integrins, which mediate cellular connection to the extracellular matrix, are involved in attachments to the cytoskeletal substratum. Cell-type-specific examples include platelet-specific integrin ($a_{IIb}b_3$), leukocyte-specific b_2 integrins, late-activation (a_Lb_2) lymphocyte antigens, retinal ganglion axon integrin (a_6b_1) and keratinocyte integrin (a_5b_1). At least 20 different heterodimer integrin receptors were known in 1998.

The cadherin molecular family of 723-748-residue transmembrane proteins provides yet another avenue of cell-cell adhesion that is cell-specific. Cadherins are linked to the cytoskeleton. The classical cadherins include E- (epithelial), N- (neural or A-CAM), and P- (placental) cadherin, but in 1998 at least 12 different members of the family were known. They are concentrated (though not exclusively found) at cell-cell junctions on the cell surface and appear to be crucial for maintaining multicellular architecture. Cells adhere preferentially to other cells that express the identical cadherin type. Liver hepatocytes express only E-; mesenchymal lung cells, optic axons and neuroepithelial cells express only N-; epithelial lung cells express both E- and P-cadherins. Members of the cadherin family also are distributed in different spatiotemporal patterns in embryos, with the expression of cadherin types changing dynamically as the cells differentiate.

OVERALL CELLULAR STRUCTURE

As recently as the late 1970s, the cell was often incorrectly described as a water-filled membranous sac enclosing a loosely structured population of discrete organelles. In reality, the interior of a cell is extremely compact, only a few times more open than a hydrated protein crystal. The cell interior is crisscrossed by many tens of thousands of cytoskeletal filaments of various gauges, attaching organelles to the nucleus, the plasma membrane, the ECM, and to each other.

Cells also have various shapes. Most cells are surrounded by and are anchored to neighbouring cells. Such immobilized cells usually assume a polyhedral shape. Specialized cells adopt shapes related to the specific functions they perform. A dramatic example is the nerve cell, which has one or more cylindrical processes extending from the cell like the branches of a tree, which processes allow the cell to receive and to transmit electrical signals.

From the nanomedical perspective, the cell may be regarded as a large machine constructed of many smaller machines. These smaller machines are the organelles. Organelles are typically 0.53 microns in diameter, roughly the same size as the largest bloodborne medical nanorobots.

Cell Membrane

The cell and most organelles are individually surrounded by their own thin envelope. The envelope surrounding the entire cell is called the plasma membrane. According to the fluid mosaic model, all membranes are composed of a double layer of lipid molecules, called the lipid bilayer, in which proteins are embedded. The lipid bilayer acts as a barrier to the diffusion of polar solutes, whereas the embedded proteins provide the pathways for

1. The selective transfer of certain molecular substances through the lipid barrier, and

2. The mechanical transfer of information from the ECM into the interior of the cell.

The plasma membrane actually represents only a tiny fraction of the total membrane surface in the cell. The combined membrane surface area of the endoplasmic reticulum is 44 times larger than the plasma membrane surface area for a typical human cell.

Lipid bilayer plasma membranes are 6-10 nm thick. The major membrane lipids are phospholipids, fatty acid chains in the range of 16-18 carbons long; chains with fewer than 12 carbons cannot form a stable bilayer. Phospholipid chains are amphipathic molecules — one end, the head, has a negatively-charged region, while the remainder of the molecule, the tail, consists of two long fatty acid chains. The phospholipids in cell membranes self-organize into a bimolecular layer, with the nonpolar fatty acid chains in the middle. The Polar Regions are oriented toward the membrane surfaces due to their attraction to the polar water molecules in the extracellular and cytosolic fluids.

The plasma membrane also contains other lipids. Cholesterol, a steroid lipid, acts as a "mortar" that fills in small gaps in the phospholipid structure, thus improving membrane impermeability to small water-soluble molecules like glucose by a factor of ten. Cholesterol also acts as a membrane antifreeze agent, decreasing bilayer fluidity at higher temperatures and preventing hydrocarbon chains of phospholipids from aggregating at lower temperatures. Plasma membranes may contain up to ~1 cholesterol molecule for each phospholipid molecule. The precise lipid composition of plasma membranes varies from one cell type to another, and also varies among the membranes of organelles within each cell type. Medical nanorobots equipped with suitable chemosensors may access this information, both for cell type identification during extracellular navigation and for organelle type identification during intracellular navigation.

Cytosol

In traditional cell biology, the "cytoplasm" is the filling substance in the large space enclosed by the plasma membrane and surrounding the cell nucleus. It includes all non-nuclear cell organelles plus all fluid surrounding these organelles. Every point inside the cell, except the nucleus, is considered part of the cytoplasm. The "cytosol" is the liquid that fills the entire cytoplasmic region, except for fluids filling the interior of the cell organelles. Finally, the term "intracellular fluid" refers to all of the fluid inside a cell — the cytosol plus the fluid inside each organelle, including the nucleus.

Fulton points out that in an average protein crystal ~40 Percent of crystal mass is solvent water — half strongly absorbed by the protein in the hydration shell and half resembling bulk water in its properties. Protein crystals may range from 20-90 Percent solvent, or 10-80 Percent protein, by weight. Muscle cells are ~23 Percent protein, red cells ~35 Percent, and actively growing cells contain 17-26 Percent protein by weight. Thus the cytosolic environment more closely resembles proteinaceous crystals than dilute solutions.

Hydration water is not immobile water like ice, but it does have reduced mobility, different solvating characteristics, a higher heat capacity, and is generally more ordered than bulk water. Water of hydration coats the macromolecular cell components such as carbohydrates, nucleic acids, and the proteins of membranes, and is required for enzyme activity. Mammalian cells can maintain glycolysis and normal respiration down to 60 Percent dehydration, which suggests that in a typical cell the cytosol may consist of ~60 Percent bulk water and ~40 Percent water of hydration. Distribution of ordered water in the cell is a heterogeneous and dynamic process, possibly cytonavigationally useful. For instance, ~55 Percent of the water of the vegetal pole region of the frog oocyte is bound water, with only ~25 Percent bound near the animal pole cytoplasm and ~10 Percent bound in the nucleus.

Minton showed that the volume occupied by proteins affects the activity of the other proteins in solution by "crowding" them into a smaller volume where they have less freedom of movement, forcing them into compact configurations that would occur far less frequently in dilute solutions. These effects can produce 50-100-fold excesses in protein activity over the values predicted from dilute solutions. Crowding also reduces diffusion mobility – 60 Percent-70 Percent of glycolytic enzymes are freely diffusing in squid axoplasm that contains 2 Percent protein by weight, but only 20 Percent of cytoplasmic proteins are freely diffusing in oocyte cytoplasm that contains 30 Percent-40 Percent protein by weight. Crowding effects could be nonspecific and due simply to protein number density, or they could be specific and maintained by selection. Mc Conkey estimates that about half of the polypeptides in the cell may participate in specific associational structures. Many enzymes exist in complexes that may involve a dozen or more proteins, potentially complicating required nanorobotic actions; according to the substrate channelling hypothesis, these complexes can help speed up reactions.

Even without these crowding effects, cells display localized biochemical gradients that are cytonavigationally significant. Specific molecules, organelles and physiological processes may be localized to defined zones within the cell. This regional cytoplasmic differentiation is regulated in part by cytosolic Ca^{++} ion and by H^+ ion (pH) spatial gradients. As another example, the cytosol immediately surrounding the Golgi apparatus is compositionally distinct from the cytosol that closely encircles the cell nucleus. Active chemical processing taking place within most cellular organelles produces persistent concentration gradients that may be used by in cyto nanorobots to establish proximity to specific organelles. Microsecond-cycle chemical nanosensors can sample the local environment far faster than the typical time required for a molecule to diffuse a distance of ~1 micron, the mean separation between adjacent major organelles inside human cells.

Control proteins such as the transcription factors that regulate gene expression are typically present in only ~300-3000 copies of each type per cell. More common cytosolic proteins may be present in numbers up to 10^6-10^7 copies of each type per cell.

RIBOSOMES

Ribosomes are the smallest and most numerous "organelle" in the human cell. A typical liver cell contains 10^7 ribosomes constituting ~5 Percent of the total dry mass of the cell. Cells less actively involved in protein synthesis have correspondingly fewer ribosomes. Some ribosomes float freely in the cytosol, making proteins for intracellular use. Others are attached to membranes or to the cytoskeleton, synthesizing proteins destined for membranes or for export from the cell.

Ribosomes are ~25 nm in diameter. Each of the several types of ribosome is constructed of two subunits that fit snugly together. A typical ribosome might have a mass of 4.2 million daltons, comprised of 2.8 million daltons for the large ribonucleoprotein 60S subunit (~23 nm diameter) and 1.4 million daltons for the small ribonucleoprotein 40S subunit (~9 nm diameter). There is no intrinsic difference between free cytosolic ribosomes and membrane-bound ribosomes – it is the "signal" sequence on the end of the protein being synthesized that directs a free ribosome to become a bound ribosome.

The ribosome is an ATP-powered protein-assembling machine. Amino acids drawn from the cytosol are presented to the ribosome, which incorporates them one by one into polypeptides at ~20 Hz. The complete synthesis of an average-sized protein takes 20-60 seconds. Even during this short period, multiple initiations take place, with a new ribosome hopping onto the 5' starting end of an mRNA molecule almost as soon as the preceding ribosome has translated enough of the amino acid sequence to get out of the way, thus allowing the assembly of many copies of the same protein to proceed almost in parallel. Thus, under physiological conditions, actively translated mRNA is found in polyribosomes, or polysomes, formed by gangs of multiple ribosomes spaced as close as 80 nucleotides apart along a single messenger molecule.

ENDOPLASMIC RETICULUM

The endoplasmic reticulum (ER), the principal protein manufacturing facility of the cell, is the most spatially extensive cytoplasmic organelle. The ER is an interconnecting membranous network of fluid-filled vesicles, branching tubules, and flattened sacs or cavities called cisternae, possessed by almost all eukaryotic cells. The 5-6 nm thick membrane of the endoplasmic reticulum typically constitutes more than half of the total membrane surface area present in a cell. The membranes enclose a space that is continuous throughout the network and connects with the perinuclear space between the two membranes of the nuclear envelope, allowing transport of manufactured substances throughout the cell. The fluid occupying the luminal or cisternal space of the ER, which is typically 20-40 nm wide but ranges from 10-70 nm, is called the reticuloplasm. There are ER-specific luminal proteins called reticuloplasmins found only in the reticuloplasm— if these proteins are detected, a nanorobot has localized its position uniquely to this space. Molecular chaperones that assist in protein folding and are unique to the ER lumen include immunoglobin heavychain binding protein (BiP) and calnexin. BiP bears the "KDEL" amino acid sequence (Lys-Asp-Glu-Leu) at its C-terminus, a tagging sequence specifying that a protein is intended for attachment to the luminal surface of the ER cisterna. There is evidence that specific compartments or subdomains within the ER may perform specialized functions and may also bear unique chemical markers, thus allowing nanorobots to distinguish different regions of the ER.

There are two varieties of endoplasmic reticulum — rough or "granular" ER, and smooth or "agranular" ER, both present in the same cell. One variety or the other may predominate in a given cell type depending on the specific functions performed by the cell, and the mix may change within the same cell during different periods of cell activity. Both are anchored to and mobilized by cytoskeletal motor proteins.

Rough ER is an extensive membranous network of flattened sheets with a flattened sac appearance and a continuous cisternal space that communicates directly with the perinuclear space. Ribosomes are bound to the ER membrane only on the cytosolic surface, loosely attached by short cylindrical conduits comprised of the Sec61 trimeric protein complex spanning the ER membrane and produced exclusively in the granular ER. Ribosome-produced proteins enter the cisternal space through these conduits, called translocons, in the ER membrane. Found in all nucleated cells except sperm, rough ER synthesizes proteins and packages proteins destined to be secreted by cells. The rough ER is also the site of the initial steps in the addition of sugar groups to glycoproteins by dolichol, a hydrophobic lipid that resides in the ER membrane with its active group facing the ER lumen. Rough ER is especially abundant in cells that are specialized for protein secretion such as the pancreatic acinar cells and antibody-secreting plasma cells, or that are specialized for extensive membrane synthesis such as the immature egg cell or the retinal rod cell. In such cases, almost half of the total ribosomes present in the cell may be bound to the granular ER. Glycoproteins produced in the rough ER become encapsulated by bits of ER membrane, which bud off and migrate along cytoskeletal elements to the Golgi complex for further processing.

Smooth ER has a branched, tubular structure with no ribosomes on its surface, though its interior is continuous with the granular ER. Agranular ER has enzymes in the cisternal side of its membrane that allow it to synthesize carbohydrates, lipids including neutral fats, phospholipids and steroids, and lipoproteins in liver cells, and that allow it to assist in drug detoxification in the liver and kidney. Muscle cells have a specialized, elaborate smooth ER called the sarcoplasmic reticulum that sequesters Ca^{++} which controls muscle contraction from the cytosol. Cells in the testes that synthesize steroid hormones from cholesterol have an expanded smooth ER compartment to accommodate enzymes needed to synthesize cholesterol and then to modify it to make steroid hormones.

GOLGI COMPLEX

The Golgi complex is a multicisternal membranous structure crudely similar to the rough ER. Substances synthesized by the ER usually pass via vesicles traversing the cytoplasm on microtubule tracks to the Golgi complex, where they are further processed, concentrated, sorted, containerized in appropriate self-identifying new vesicles, and then shipped out to various destinations. Some oligosaccharides are synthesized in the ER, while extensive portions of the same structures are degraded in the Golgi – competing sets of reactions that are crucial in fixing the "address" of the intracellular compartment to which a newly synthesized macromolecule will be sent. Thus the GC serves as a production editor, allowing several successive sorting steps to take place while reducing the overall cytomanufacturing error rate. The Golgi acts as a "countercurrent" fractionation system to separate proteins destined for the plasma membrane from those to be retained in the ER.

Morphologically, the Golgi complex consists of a series of flattened, membranous saccules, forming disk-shaped cisternae that are stacked together to form a cup-shaped structure. Cisternal boundaries are morphologically distinct and are only rarely, if at all, connected. A series of 5-8 saccules comprises a single Golgi stack or dictyosome, typically measuring ~1 micron in diameter and ~250 nm thick. As with the ER, the size and number of Golgi complexes vary according to cell type and metabolic activity. All eukaryotic cells have a GC. Some cells have just one stack; others, particularly those especially active in secretion, may have hundreds. One enzyme, thiamine pyrophosphatase, occurs only in Golgi membranes, and galactosyl transferase is found only in the GC. These may be used by nanorobots as unique cytochemical markers of the Golgi complex.

The Golgi complex is a dynamic structure. The saccules of the GC form by the fusion of vesicles budded off from the ER. As these vesicles reach the GC and fuse, they give rise to new saccules on the forming face of the organelle. Existing saccules move forward (outward) in the stack. Meanwhile, on the maturing face of the GC, vesicles bud off the tips of the saccules continuously and exit the complex, transported on cytoskeletal elements, through an interaction with motor proteins. This is known as the "cisternal maturation model." Within the Golgi, net membrane flow can be quite rapid. In certain mucus-secreting cells of the intestinal mucosa, it takes only ~2000 sec for membrane components to move from the forming face of a Golgi stack to the maturing face. Vesicles exhibiting COPI (coat protein I) apparently transport some proteins backwards, both within the Golgi stack and also from the Golgi to the ER. One study counted ~400 individual vesicles in transit within ~0.2 microns of a ~4 micron^3 region containing 2 stacks of 7 cisternae.

The two faces of a Golgi stack are biochemically distinct. Specific enzymes and receptor proteins are concentrated in the cisterna on the "cis" face of the stack (the "receiving" end), while other proteins are localized mainly in the cisterna on the "trans" face (the "shipping" end). The receptor protein for the carbohydrate mannose-6-phosphate (M6P) is present only in the forming cisterna of the Golgi stack, thus allowing the stack to recognize these proteins and target them to the lysosomes. A 28,000-dalton protein called GS28 acts as one of the targeting receptors involved in the recognition of ER-derived vesicles that are destined for fusion with the cis-Golgi membrane. By 1998, more than two dozen different members of the small G protein superfamily (including the Rab and Arf protein families) had been implicated in the regulation of intracellular vesicular trafficking and membrane recognition. The Golgi complex also contains substantial amounts of cholesterol and sphingolipids, making a rising gradient from cis to trans. Various Golgi compartments are defined by the presence of particular enzymes that are localized to one or more compartments, often in a graded manner. While all cisternae are fenestrated and display coated buds, the trans-most cisterna produces exclusively clathrin-coated buds, whereas the others display only nonclathrin coated buds. Such distinctions will allow in cyto medical nanorobots with suitable transmembrane chemical sensor tools to readily establish Golgi stack configuration, polarity, and physical extent.

The Golgi stack has other easily detectable asymmetries. Since the GC mediates the flow of secretory proteins from the ER to the cell exterior, the GC displays spatially variant protein and lipid compositional gradients along the Golgi polarity axis. Specifically, the saccule membranes at the forming face resemble those of the ER in morphology and composition — ~5-6 nm thick membranes defining a relatively narrow ~30-80 nm cisternal space, with ~20 Percent phosphatidylcholine in the membrane. At the maturing pole, the saccule membranes more closely resemble the plasma membrane, ~10 nm thick, with a much wider cisternal space (>100 nm) and with ~10 Percent phosphatidylcholine in the membrane. (Variable membrane thickness is part of the control mechanism for spatially anisotropic budding). Another asymmetry: Swarms of ~50 nm diameter membrane-bound vesicles always cluster on the side of the Golgi stack abutting the ER and along the circumference of a stack near the dilated rims of each cisterna.

MITOCHONDRIA

Mitochondria are the principal chemical energy transducers of the eukaryotic cell under aerobic conditions. Except for the 10 reactions of the glycolytic pathway, all of the ATP-generating capacity of eukaryotes lies within the mitochondria. Mitochondria are scattered throughout the cytoplasm in virtually all aerobic cells, with numbers ranging from ~300/cell for relatively inactive cells like lymphocytes up to 2000-3000/cell for very active cells such as liver, kidney tubule and cardiac muscle cells. Mitochondria are often clustered within the cell in regions of intense metabolic demand. In muscle cells the mitochondria are organized in rows between adjacent contractive myofibrils in order to minimize the required diffusion distance for ATP molecules that are powering the activity. Similar localization appears in sperm tail flagella, in cilia, and at the base of kidney tubule cells where exchange with the blood is most rapid. Except for plant chloroplasts, mitochondria are unique among organelles in having their own DNA genomes, ribosomes and tRNAs that are quite different from those found in the cytoplasm.

After the nucleus, the mitochondrion is the largest organelle in most animal cells. The typical time-averaged dimensions are roughly cylindrical, with a 0.5-1.0 micron diameter and a ~3 micron length — about the size of a bacterium or modest-sized medical nanodevice. Mitochondria are usually depicted as tiny kidney-bean or sausage-shaped organelles, but in living cells they squirm, flex, elongate, and change shape almost continuously in part due to cytoskeletal interactions. Mitochondria may be shuttled around a cell at up to ~10 microns/sec via dynein motors riding on microtubules, bulging the plasma membrane as they travel.

Mitochondria have four functional compartments: outer membrane, intermembrane space, inner membrane, and the matrix. The outer surface is a smooth, featureless, 5-6 nm thick membrane embedded with many copies of a transport protein that forms large aqueous channels through the lipid bilayer, so the membrane is permeable to all molecules of mass <10,000 daltons. Since mitochondria do not grow by fusion with vesicles synthesized elsewhere in the cell, transfer of phospholipids from the endoplasmic reticulum to the mitochondrion requires special phospholipid transfer proteins in the outer membrane that can recognize a specific kind of phospholipid, remove it from a vesicle membrane and add it to the mitochondrial membrane. Special enzymes such as monoamine oxidase, acyl-CoA synthetase, and phospholipase A_2 are also present. The first 32 N-terminal amino acids of cytochrome c1 constitute a fixed "leader sequence" that is recognized as a matrix targeting signal and allows the tagged molecule admission through the outer membrane. Thus, as with other organelles, the outer coat of the mitochondrion should be immediately and uniquely recognizable by in cyto nanorobots equipped with suitable chemosensors.

The intermembrane space also contains numerous unique proteins, as the essential multispanning carrier proteins Tim10p and Tim12p.

The 5-6 nm thick inner membrane is highly convoluted, forming a series of ~30 infoldings (each fold ~100 nm wide) known as cristae, which extend into the inner compartment or matrix, roughly quintupling the active surface area of the outer membrane. The inner membrane is rich in cardiolipin, a phospholipid that accounts for 10 Percent of the membrane lipid content and renders it unusually impermeable to most solutes. The inner membrane also contains a variety of special transport proteins that make it selectively permeable to those particular small molecules that are metabolized by mitochondrial enzymes concentrated in the matrix space, such as the integral membrane proteins Tim17, Tim 23, and Tim22p. All of these membrane-specific proteins are readily detectable by nanorobots that are extending manipulators tipped with appropriately configured chemical sensors through the intermembrane space.

The inner membrane is the locale of electron transport and ATP synthesis. A single mitochondrion has ~10,000 large protein F_1 (ATP synthase) complexes embedded in its inner membrane, randomly distributed and typically measuring ~10-20 nm in diameter. These integral proteins freely diffuse laterally within the inner membrane. Mobility is high – protein complexes collected at one end of a membrane by externally imposed electrophoretic forces return to a random distribution in just a few seconds.

The prominence of cristae is correlated with the relative metabolic activity of the cell or tissue. Heart, kidney, and muscle cells have high respiratory activity levels, and their mitochondria have correspondingly large numbers of prominent cristae. The number of cristae is three times greater in the mitochondria of cardiac muscle cells than in hepatic mitochondria, reflecting the greater demand for ATP in heart tissue. The cristae of mitochondria in different cell types are not only different in number and deepness, but also in basic morphology.

The interior of the mitochondrion is filled with a semifluid gel-like matrix that contains a concentrated mixture of hundreds of different enzymes related to aerobic cellular respiration and oxidation, plus several identical copies of the mitochondrial DNA genome, special mitochondrial ribosomes, tRNAs, and various enzymes required to express the mitochondrial genes. The mitochondrial genome consists of a circular DNA molecule of ~11 million daltons, coding for about a dozen polypeptides. It has 16,569 base pairs and a contour length of ~5 microns. Mitochondria are self-replicating organelles (with the help of cell-supplied proteins and lipids). When cellular requirements for ATP increase, mitochondria pinch in half to increase their number, then both halves regrow to the former full size.

Mitochondria also are sometimes found as extended reticular networks which are extremely dynamic in growing cells, with tubular sections dividing in half, branching, and fusing to create a fluid tubular web.

NUCLEOGRAPHY

Nucleography is the "geography" of the cell nucleus. The nucleus, 5-8 microns in diameter for a 20 micron tissue cell and up to 10 microns for a fibroblast cell, is the largest cellular organelle and the only one that is voluminous enough, in theory, to admit a micron-scale medical nanorobot into its interior. The nucleus is usually a large spherical or ovoid structure surrounded by its own nuclear membrane, although its shape generally conforms to the shape of the cell. If a cell is elongated, the nucleus may be extended as well.

Almost all cells contain a single nucleus, whose primary function is the storage and expression of genetic information. However, a few cell types have multiple nuclei of similar size, such as skeletal muscle cells, osteoclasts, megakaryocytes, and some hepatocytes. A few cell types have no nucleus, such as red blood cells, platelets, keratinized squamous epidermal cells, and lens fibres.

In 1998 the finer details of nuclear structure were just beginning to be understood. Thus the following discussion of nucleography must be regarded as a very tentative work in progress.

NUCLEAR ENVELOPE

The nuclear envelope enclosing the nucleus is a lipid bilayer similar in structure to the cell membrane, except that it is a double-layered membrane which is topologically more convenient for dissolution during mitosis and subsequent reassembly from vesicles. Each of the two lipid bilayer membranes is 7-8 nm thick. The outer nuclear membrane is occasionally continuous with the rough endoplasmic reticulum and is almost entirely surrounded by it. Like the rough ER, the outer membrane is often studded on its outer surface with ribosomes involved in protein synthesis. Intermediate filaments extend outward from the outer membrane into the cytoplasm, anchored on the other end to the plasma membrane or other organelles, thus positioning the nucleus firmly within the cell and increasing its mechanical stiffness almost tenfold.

The perinuclear space between the two membranes ranges in width from 10-70 nm but is usually a gap of 20-40 nm. This fluid-filled compartment is continuous with the cisternae of the rough ER, thus providing one possible avenue for transporting substances between the nucleus and different parts of the cytoplasmic compartment.

The nuclear envelope disassembles at the onset of mitosis and is reassembled at the end of mitosis.

NUCLEAR PORE COMPLEXES

The most distinctive feature of the nuclear envelope is the presence of numerous nuclear pores, small cylindrical channels with eightfold symmetry that extend through both membranes and provide direct contact between cytoplasm and nucleoplasm. Each pore complex marks a point of fusion between the inner and outer membranes. Elements of the cytoskeleton appear to be attached to many pores, possibly allowing direct mechanical regulation of pore activity.

Each nuclear pore complex is a huge multimolecular assemblage measuring 70-90 nm in diameter, with a mass of

125 million daltons, ~34 times the size of a ribosome. Up to 100 different nucleoporin protein molecules make up the structure. Early experiments with passive gold particles showed that cytoplasmic particles with diameters of 5-6 nm passed into the nucleus in ~200 sec, those with diameters of 9-10 nm took ~10^4 sec, but particles larger than 15 nm didn't seem to enter at all. Closer examination has revealed that the pores are actually large enough to allow the passage of substrates as large as 23-26 nm, but this is still much too narrow for nanorobots or their flexible processes to pass through without damaging the mechanism. The nuclear localization sequence, a molecular tag consisting of 1-2 short sequences of amino acids, marks cytoplasmic proteins for active transport through the nuclear pores. Small (~40 nm) arm-like import receptors ringing the mouth of the pore bind to a protein cargo tagged with an NLS, then flex toward the pore to shove the cargo into the mouth.

The density of pores across the surface of the nuclear envelope varies greatly, depending mainly on cell type and the amount of RNA being exported to the cytoplasm. Values range from 3-4 pores/micron^2 in some white cells up to 50 pores/micron^2 in oocytes and a theoretical maximum density of 60 pores/micron^2. A typical ~20 micron human cell has 2000-4000 pores embedded in its nuclear surface, a mean density of 10-20 pores/micron^2. Pore structures may protrude at most ~100 nm into the nucleoplasmic space.

Nuclear Cortex

The nuclear cortex is an electron-dense layer of intermediate filaments on the nucleoplasmic side of the inner nuclear membrane. The cortex, also called the nuclear lamina or karyoskeleton, is up to 30-40 nm thick in some cells but is difficult to detect in others. Its proteinaceous fibres are arranged in whorls that may serve to funnel materials to the nuclear pores for export to the cytoplasm. These fibres may also be involved in pore formation. The nuclear cortex helps to determine nuclear shape, and also binds to specific sites on chromatin, thereby guiding the interactions of chromatin with the nuclear envelope. Chromatin binding sites on the nuclear cortex avoid the immediate vicinity of nuclear pores to ensure unobstructed passage of materials through the pores.

Nucleoplasm and Chromatin

The nucleoplasm is the semifluid matrix in the interior of the nucleus. It contains some condensed but mostly extended chromatin (called heterochromatin and euchromatin, respectively), as well as a structural nuclear matrix of nonchromatin material. The chromatin represents chromosomes as they exist between cell divisions. Chromosomes assume a highly condensed state as the cell prepares to divide, but after mitosis most of the chromosomes relax into a highly extended state. The nucleus of the human cell contains 46 chromosomes of varying lengths, in 23 pairs. Each of these in turn are composed principally of a single deoxyribonucleic acid (DNA) molecule. The DNA contains the genes of the cell, and all ~100,000 genes are represented, though not expressed, in each nucleated cell. The nucleosol, or fluid component of the nucleoplasm, contains salts, nutrients, and other needed biochemicals. A number of different granules are also present.

During interphase (e.g., between cell divisions), individual chromosomes occupy compact, discrete territories within the nucleus that may range up to 4 microns in diameter. The structure and location of these territories is specific for both cell type and mitotic stage, and may be arranged in the same spatial order as is found in the wheel-shaped ring aggregate known as the chromosome rosette at the time of mitotic prometaphase. It has been proposed that active genes are preferentially localized to the periphery of the chromosome territories; RNA templates would be preferentially produced at the surfaces of these territories and then shed into interchromosomal domain channels for further processing and transport. Others have argued against this. But knowledge of any such spatial orderings, which was woefully incomplete in 1998, might permit intranuclear nanorobots to approximate their position inside the nucleus once the locations of three or more specific chomosomes have been definitively established. The diameter of a territory $D_{territory} \sim D_{nucleus} (c_{chromosome} / c_{genome})^{1/3} \sim 3$ microns, where $c_{chromosome}$ is chromosome size and c_{genome} is genome size, both measured in base pairs, and $D_{nucleus}$ is the diameter of the nucleus.

Note, however, that these territories are not rigid. Changes in the relative positions of chromosomal territories often occur at ~0.3-0.4 nm/sec, and intraterritorial movement and flexing of subchromosomal foci measuring 400-800 nm in diameter have also been observed. Both cytoplasm and nucleoplasm may contain numerous as yet undiscovered intricate substructures that could provide many new navigational aids. An important task for early nanomedicine-oriented research will be to fully explore and elucidate this fine structure, and to determine whether or not it is stable enough to be relied upon in any way for intracellular navigation.

In its most relaxed state, chromatin resembles a network of bumpy threads weaving their way through the nucleoplasm. Chromatin is composed of roughly equal amounts of negatively charged DNA (comprising the chromosomes) and globular histone proteins. Nucleosomes, the fundamental units of chromatin, are spherical clusters of eight histone proteins, connected like beads on a string by a DNA molecule that winds around each of them. The average cell nucleus contains 25 million nucleosomes, also called histone octamers. Each nucleosome is encircled by 146 base pairs of DNA. Nucleosomes have a mass of 206,000 daltons. About half of nucleosome mass is protein and half is DNA. Each human chromosome contains DNA with an average contour length of ~75 mm.

NUCLEOLUS

The largest and most prominent "nuclear organelle" is the nucleolus, a highly coiled structure associated with numerous particles but not surrounded by a membrane. The nucleolus is a ribosome-manufacturing machine. Assembly of precursor ribosomal subunits within the nucleolus requires ~1800 sec, while the complete assembly of a large ribosomal subunit (needing only protein to make a completed ribosome) takes ~3600 sec.

The nucleolus is composed of DNA, RNA, and proteins. It also has a granular component and a fibrillar component, and a variable internal structure. The granular component consists of ~15-nm particles that are ribosomal subunits in the process of maturation. The fibrillar component consists of rRNA molecules that have already become associated with proteins to form fibrils with a thickness of ~5 nm. The size of the nucleolus correlates with its level of activity. In cells characterized by a high rate of protein synthesis and hence by the need for many ribosomes, the nucleolus can occupy 20-25 Percent of nuclear volume, mostly comprised of the granular component. In less active cells, the nucleolus is much smaller — as small as 0.5 micron in a mature lymphocyte. Nucleoli are frequently located at or near the nuclear envelope, adhering directly to the nuclear lamina or attaching to it by a pedicle. In nuclei having a centrally located nucleolus, the nuclear envelope is folded to form a nucleolar canal that is in direct contact with the nucleolus.

Most human nuclei contain only one nucleolus, except for liver cell nuclei which may contain more than one nucleolus and cultured HeLa (cancer) cells which may have up to six. The number of nucleoli in a eukaryotic cell nucleus normally is determined by the number of chromosomes with secondary constrictions, or nucleolus organizer regions. The human genome contains five NORs per haploid chromosome set, or 10 NORs per diploid nucleus, each located near the tip of a chromosome. However, instead of 10 separate nucleoli, the typical human nucleus contains a single large nucleolus representing the fusion of loops of chromatin from the 10 separate chromosomes with NORs. The DNA from the remaining diploid chromosomes is distributed in specific regions throughout the nucleoplasm. During mitosis, the chromosomes condense into a more compact form and the nucleolus shrinks, then disappears altogether. A cell undergoing mitosis thus has no nucleolus and synthesizes no rRNA. Once mitosis is complete, the nucleolus reappears. As rRNA synthesis resumes, ten tiny nucleoli appear, one near the end of each of 10 different chromosomes; these enlarge, eventually fusing into the single large nucleolus characteristic of the interphase human nucleus.

8

Locomotion

NANOROBOT DEXTERITY AND MOBILITY

Manipulation and mobility are crucial basic capabilities in most classes of medical nanodevices. Manipulation includes handling fluids, biological objects such as tissue matrix fibres or cellular elements, and nanomachines or their components. Physicians must be able to direct tissue or cell-repair nanorobots to travel to a specific site where treatment is required, and once there, to manipulate the local environment to achieve the desired results. Nanodevice mobility in vivo makes possible the rapid reconfiguration of nanomedical communication, navigation, and power systems, while ex vivo mobility allows the design of more robust diagnostic and personal defensive systems. In vivo locomotion also permits precise mapping of the internal regions of the human body across many size and time scales, both for diagnostic and for therapeutic purposes.

Adhesion and Fluid Transport

An understanding of surface forces is an essential preliminary to any study of mechanisms of manipulation or locomotion. Such adhesive forces become important at the submicron scale where nanorobotic parts are meshing with and sliding against one another. Surface adhesion forces may cause tools, parts or workpieces to stick together. Surface forces may cause airborne or solvent-immersed nanodevices to adhere to container walls, to other dry or humid surfaces, or to each other possibly causing clumping. For a 1-micron object, the adhesive forces may exceed gravitational and inertial forces by a factor of 10^6 or more. Capillary forces also are important in fluid transfers inside nanodevices, in fluid transfers between nanodevices or between nanodevices and their operating environment, and during nanodevice tasks requiring locomotion through, or limb/object manipulation in, a watery environment. Surface tension forces are especially relevant during locomotion missions requiring passage through air/water interfaces, as might occur in the lungs.

Bowling classifies adhesive forces between objects into three categories. The first category includes long-range attractive interactions that may bring a particle to a surface and establish the adhesion contact area, such as van der Waals forces, electrostatic forces, and magnetic forces. A second category of forces are the interfacial reactions that help to define the adhesion area, most importantly the capillary forces arising from the establishment of liquid or solid bridges between particle and surface, but also including other effects such as sintering, diffusive mixing, mutual dissolution and surface alloying, which will not be considered further here. In the third category are very short-range interactions that may strengthen adhesion after an adhesive contact area has already formed; such forces include chemical bonds of all types and various noncovalent bonds such as hydrogen bonds that have already been described.

Van der Waals Adhesion Forces

In dry or vacuum environments, adhesion forces between <100 micron diameter particles and surfaces (nanorobot and other) separated by 100 nm or less are usually dominated by van der Waals interactions. Consider a spherical particle of radius r lying a small distance $z_{sep} <\sim 100$ nm from a flat plate of surface area $A >> p\, r^2$. As a crude approximation, ignoring retardation effects that may become important in solution for $z_{sep} > 5$ nm, the van der Waals adhesive force is approximated by:

$$F_{vdW} = \frac{Hr}{6 z_{sep}^2}$$

for z_{sep} << r, where H is the Hamaker constant for the interaction. The Hamaker constants for the materials comprising the sphere, the plate, and the surrounding medium are H_s, H_p, and H_m, respectively, giving:

$$H - \left(H_s^{1/2} - H_m^{1/2}\right)\left(H_p^{1/2} - H_m^{1/2}\right)$$

The adhesion separation distance z_{sep} is often taken as ~0.4 nm for particles in intimate contact with a surface. Hence the force required to overcome the van der Waals adhesive attraction of a perfectly rigid r=0.5 micron particle adhered to a diamond plate in vacuo (with z_{sep}=0.4 nm, H=340 zJ) is F_{vdW} ~ 180 nN.

Real particles are not perfectly rigid but deform under the influence of this adhesive force. If the adhesion surface area is increased from point contact to a circle of radius $r_{adhesion}$, then the van der Waals adhesion force increases to:

$$F_{vdw} = \left(\frac{H r}{6 z_{sep}^2}\right) + \left(\frac{H r_{adhesion}}{6 z_{sep}^3}\right)$$

According to the JKR theory of adhesion mechanics for a sphere on a flat surface of the same material with a work of adhesion $W_{adhesion}$ and elastic modulus K_e:

$$r_{adhesion} = \left(\frac{6\pi r^2 W_{adhesion}}{K_e}\right)^{1/3}$$

Assuming $W_{adhesion}$ ~10 J/m^2 and K_e ~ 10^{12} N/m^2 for smooth flawless diamond and r=0.5 micron gives $r_{adhesion}$ ~ 36 nm. Taking z_{sep}=0.4 nm contact with a flat diamond plate over a ~4100 nm^2 circular adhesion spot in vacuo, then F_{vdW} ~ 1300 nN, a mean surface pressure of $p_{mean}=F_{vdW} / \pi r_{adhesion}^2$ ~ 3200 atm over the contact circle and a peak pressure p_{peak}=1.5 p_{mean}=4800 atm at the centre. As the sphere is pulled away, separation occurs abruptly when the contact radius falls to $r_{adhesion}$ / $4^{1/3}$=(~63 Percent) $r_{adhesion}$=23 nm.

Van der Waals forces also depend upon the roughness of the surface. If a smooth sphere approaches a surface having uniform rugosity or roughness b_{rug} with roughness feature size << r, then the van der Waals force is approximately:

$$F_{rug} = \left(\frac{z_{sep}}{z_{sep} + \frac{1}{2} b_{rug}} \right) F_{vdW}$$

For z_{sep}=0.4 nm and b_{rug}=10 nm, F_{rug}=2 Percent F_{vdW} ~ 4 nN for the rigid sphere in the examples given above. Surface dimpling is a common practice in MEMS fabrication, to reduce adhesion between polysilicon layers and the substrate.

A few other geometries of rigid bodies approaching contact are useful as well. The van der Waals adhesive force between two flat plates of equal interfacial area A and uniform separation z_{sep} is:

$$F_{vdW} = \frac{H\,A}{24\pi\, z_{sep}^3}$$

Two (1 nm)^2 diamond plates separated by z_{sep}=0.4 nm in vacuo require F_{vdW} ~ 0.07 nN (~7 atm) to overcome the van der Waals attrâctive force. Such plates are the same size as the contacting end surfaces of the logic rods employed in Drexler's mechanical computer design; the design assumes a rod realignment force of 1 nN in the exemplar calculations. This result suggests that forces on the order of 10–100 pN must be applied to mobile nanoscale components inside mechanical nanodevices in order to overcome static van der Waals adhesion that arises due to the contact proximity of parts within the structure.

Two rigid spheres of radii r_1 and r_2, separated by a distance z_{sep} << r_1, r_2, have a van der Waals adhesive force of:

$$F_{vdW} = \frac{H\,r_{red}}{6\,z_{sep}^2}$$

where the reduced radius $r_{red}=(r_1 r_2) / (r_1 + r_2)$. A pair of 1-micron diameter rigid diamond spheres separated by $z_{sep}=0.4$ nm in vacuo have a van der Waals mutual adhesive force $F_{vdW} \sim 90$ nN.

As one final example, the van der Waals force between two parallel cylinders of length L, radii r_1 and r_2, and separation z_{sep} is:

$$F_{vdw} = \frac{H L r_{sep}^{1/2}}{128^{1/2} z_{sep}^{5/2}}$$

A pair of 1-nm diameter, 10-nm long rigid diamondoid cylinders lying exactly $z_{sep}=0.4$ nm apart along their entire length in vacuo feel a van der Waals force of $F_{vdW} \sim 1.5$ nN. In all three examples, as before, the force rises slightly if the objects deform and declines significantly if surface rugosity is increased.

Electrostatic Adhesion Forces

Two types of electrostatic forces may act to hold particles to surfaces. The first type of force is due to bulk excess charges present on the particle or surface which produce a classical Coulombic attraction known as the electrostatic image force. Consider two spheres of radii r_1 and r_2 located a distance z_{sep} apart, where $z_{sep} << r_1, r_2$. Then following Eqn. the image force is given by:

$$V_{HP} = \frac{\pi r_{tube}^4 \Delta p}{8 \eta l_{tube}} = \pi r_{tube}^2 v_{flow} \ (m^3/\text{sec})$$

where q_1 and q_2 are charge on the two spheres, s_1 and s_2 are the surface charge densities (coul/m^2), $e_0=8.85\times10^{-12}$ farad/m (permittivity constant), k_e is the dielectric constant of the medium, and $d_{sep} \sim r_1 + r_2$ is the distance between charge centres. Taking $s_1 \sim s_2=10^{-5}$ coul/m^2, $r_1 \sim r_2=0.5$ micron and $k_e=1$ in vacuo, then $F_{image} \sim 0.009$ nN.

The electrostatic image force between two flat plates of equal interfacial area A, equal charge density s, and uniform separation z_{sep} with $z_{sep} << A^{1/2}$, is:

$$F_{image} = \frac{\sigma^2 A}{2 \varepsilon_O \kappa_e}$$

For s ~ 10^{-5} coul/m^2 and A=1 micron^2 in vacuo (k_e=1), F_{image} ~ 0.006 nN (~0.00006 atm). At larger z_{sep} and a voltage differential V_{plate} between the plates with capacitance C_{plate}, then:

$$F_{image} = \frac{C_{plate} V_{plate}^2}{2 z_{sep}} = \frac{\varepsilon_O \kappa_e A V_{plate}^2}{2 z_{sep}^2}$$

Thus if A=1 micron^2, V_{plate}=1 volt, and z_{sep}=0.2 micron, then F_{image} ~ 0.1 nN.

The second and more important type of electrostatic force for very small particles is the electrostatic contact potentials, also known as induced electrical double layer forces. When two different initially uncharged materials with different local energy states and work functions are brought into contact, charge flows between them until an equilibrium is reached. The resulting potential difference is called a contact potential, $V_{contact}$, which typically ranges from ~0–1 volt. In the case of two metals, only the surface layer which is $x_{contact}$ ~ 1 nm thick (the gap for tunneling) carries contact charges; for semiconductors and insulators, these regions may extend into the bulk up to $x_{contact}$ ~ 1 micron or deeper. For a spherical particle of radius r at a distance z_{sep} from a surface, the induced charge density is $\sigma = \varepsilon_0 V_{contact} / z_{sep}$ and the contact force is:

$$F_{contact} = \frac{\pi \varepsilon_0 r V_{contact}^2}{z_{sep}}$$

Taking r=0.5 microns, $V_{contact}$=0.5 volts, and z_{sep}=5 nm, then $F_{contact}$ ~ 1 nN (~0.01 atm). Induced charge densities vary experimentally from 2 x 10^{-7} coul/m^2 (~$2{\times}10^{-11}$ atm contact pressure) for polystyrene on polystyrene up to $2{\times}10^{-2}$ coul/m^2 (~200 atm) for SiO_2 on mica. In this latter study, a pair of contact-adhered silica/silica sheets required a force of ~0.08 nN/nm to pull them apart and a mica/mica pair required ~ 0.11 nN/nm; however, due to contact electrification a silica/mica pair required 6.6-8.8 nN/nm to separate. Proper nanomechanical design can avoid or enhance these effects, as required.

Double layer electrostatic contact forces usually predominate over electrostatic image forces for small particles; image forces are important only for materials that can carry high surface charges, such as polymers of poor conductivity or other extreme insulators. Because of their r/z_{sep} dependency, contact electrical forces are usually more important than van der Waals forces only for large particles and at separations wider than ~100 nm, although electrical forces can dominate in smaller particles if the surface asperities of the particles are significant enough to remove the bulk of the particle from the contact point, thus greatly reducing the van der Waals interaction. In either case, the total forces of adhesion far outweigh the force of gravity, given by:

$$F_{gravity} = \frac{4}{3}\pi r^3 p g$$

For density r=3510 kg/m^3 for diamond, acceleration of gravity g=9.81 m/sec^2, and particle radius r=0.5 micron, $F_{gravity}$ ~ 10^{-5} nN; for r=1 nm, $F_{gravity}=10^{-13}$ nN.

Entropic and electrostatic effects can sometimes interact to produce an apparent repulsion between opposite charges – as, between positively charged bilayer surface and negatively charged colloidal particles.

Capillarity and Nanoscale Fluid Flow

In vivo nanodevices will often find it necessary to inject or to extract small aliquots of fluid from organelles, cells, tissues, or the environment. Internal gas or liquid transfers will also be commonplace in fluidic tethers, hydraulic communication and power conduits, nanohydraulic pistons, manipulators and metamorphic bumper drivers, and in bulk transfers into sensor cavities or chemical reaction chambers inside nanofactories. Nanopipes may also be used in biomimetic nanosystems such as artificial organs and artificial cell components such as tubulin substitutes. Hence it is essential to examine the nature of fluid flows and "wicking" in nanocapillary vessels, and the likely behaviour of fluids in nanomachines.

The theory of macroscale capillarity is well-studied. The concept of "wetting" is basic. Consider a fluid in a vessel. At the solid-liquid-gas interface there is a characteristic contact angle θ, the angle at which the meniscus contacts the container wall. This angle is indicative of the balance between the forces of (1) adhesion between the liquid and the solid wall and (2) cohesion within the surface of the liquid. A contact angle <90° means that adhesion is strong and the liquid is "wetting" the tube; an angle >90° implies that adhesion is relatively weak and the liquid is "nonwetting." In air against glass, the liquids alcohol, glycerol, and pure water have $\theta \sim 0°$, while turpentine has $\theta \sim 17°$ and impure water has $\theta \sim 25°$; all are wetting. On the other hand, mercury on glass has $\theta=140°$ and water on paraffin has $\theta=109°$; both are nonwetting.

In the classical example of capillarity, one end of a narrow-bore open-ended tube is dipped vertically into a liquid. The liquid wets the tube, and the liquid rises until the force due to surface tension (F_{cap}) pulling the liquid upward equals the force of gravity (F_{grav}) pulling the column of liquid downward at some height h_{cap}, given by:

$$h_{cap} = \frac{2\gamma_{lg}\cos(\theta)}{g\, r_{cap}\left(pl - p_g\right)}$$

with wetting coefficient $\cos(\theta)=(\gamma_{sg} - \gamma_{sl}) / \gamma_{lg}$ (the Young equation) where g is surface tension at the solid-gas, solid-liquid, or liquid-gas interfaces; ρ_l and ρ_g are liquid and gas density; g=9.81 m/sec and r_{cap} is the inside diameter of the tube. Thus pure water with air at STP in a glass capillary with r_{cap}=1 micron can rise h_{cap}=15 meters against gravity, representing a mean pressure $p_{cap}=h_{cap}\, g\, (\rho_l - \rho_g)$=1.5 atm. A nonwetting liquid does not rise up the tube, but falls instead.

In the more general case of a linear capillary tube whose flow vector makes an angle φ_{grav} with the ambient gravity field (e.g., $\varphi_{grav}=\pi$ is a tube pointing straight up), then the force on the fluid column is:

$$F_{column} = F_{cap} + F_{grav}$$

$$F_{cap} = 2\pi \gamma_{lg} \cos(\theta) r_{cap}$$

$$F_{grav} = \pi h_{cap} \left(\rho_l - \rho_g\right) g \cos\left(\varphi_{grav}\right) r_{cap}^2$$

Taking the largest reasonable tube that might fit inside an in vivo nanorobot, h_{cap}=1 micron, r_{cap}=0.1 micron, $\rho_l - \rho_g$ ~ 1000 kg/μ^3, g_{lg} ~ 72 x 10^{-3} N/m for pure water, and taking cos(q) ~ cos(φ_{grav}) ~ 1, then F_{cap}/F_{grav} ~ 10^8, giving the familiar result that gravity can usually be ignored in nanoscale systems. F_{column} ~ F_{cap}=44 nN and column fluid pressure p_{column} ~ 2 γ_{lg} cos(θ) / r=14 atm. Taking instead the smallest reasonable nanocapillary tube with h_{cap}=20 nm and r=2 nm, then F_{cap}/ F_{grav} ~ 10^{11}, F_{column} ~ 0.9 nN and p_{column} ~ 700 atm, making a quite substantial wicking force. Of course, the frictional effect of the interface region requires that a finite pressure P_0 ~ 2 α / r_{cap} must be applied before fluid motion will begin, where α=0.01-0.1 N/m (experimental); taking α ~ 0.01 N/m, then P_0 ~ 2 atm for r=0.1 micron, or ~100 atm for r=2 nm.

The capillary force may also be compared to the simple hydrostatic suction force:

$$F_{suck} = \pi r_{cap}^2 \Delta p$$

A nanopipette with inside radius r_{cap}=0.1 micron which is pressed against a plasma membrane surface and is then evacuated behind the attachment point sufficiently to create a surfacial pressure differential of Δp=0.3 atm produces a purely hydrostatic suction force of F_{suck}=1 nN.

Such classical continuum models assume, among other things, that the molecular graininess of the fluid can be ignored. This assumption fails when tube dimensions are comparable to the characteristic molecular length scale of the fluid. In a gas, λ_{gas} is the mean free path between collisions. For T_{gas}=310 K air with n_{gas}=2.4 x 10^{25} molecules/m^3 at p_{gas}=n_{gas} kT_{gas}=1 atm pressure, the free path is given by:

$$\lambda_{gas} = \left(2^{1/2} \pi\, n_{gas}\, d_{gas}^2\right)^{-1} = \left(2^{1/2} \pi\, p_{gas}\, d_{gas}^2 / kT_{gas}\right)^{-1}$$

$$\sim 200\, nm$$

under ideal gas conditions, if we take the effective molecular diameter as $d_{gas} \sim 0.2$ nm. At high pressure, van der Waals equation must be used, but for p_{gas}=1000 atm with $n_{gas}=1.3\times10^{28}$ molecules/m^3, then $\lambda_{gas} \sim 0.4$ nm. In a liquid, $\lambda_{liq} \sim d_{liq}$, the molecular diameter; for water molecules, $\lambda_{liq} \sim$ 0.3 nm.

A ratio of $r_{cap}/\lambda \sim 1$ marks an important transition because it implies that molecules are colliding with the walls about as often as they are colliding with each other. If $r_{cap} << \lambda$, then intermolecular interaction is rare, wall collisions are relatively frequent, and molecular motion is largely ballistic between wall impacts. But if $r_{cap} >> \lambda$, then molecule/wall interactions are relatively rare and intermolecular collisions are relatively frequent. In the latter situation the continuum flow relations should become valid, when $r_{cap} >> 200$ nm for gases at ~1 atm or when $r_{cap} >> 0.4$ nm for gases at ~1000 atm, or when $r_{cap} >> 0.3$ nm for liquids.

The range of non-continuum skin layer effects appears quite narrow in many cases. Two smooth mica surfaces that are slowly brought together while immersed in various nonaqueous fluids display oscillating attractive and repulsive forces as a function of separation when the gap is less than 6-10 molecular diameters; periodic jump distances are about equal to the molecular diameter of the fluid. Oscillations become smaller between rough surfaces and in liquids that mix molecules of different sizes. Similar experiments in water with electrolyte produce similar behaviour, but with ionic double layer repulsions superimposed — oscillatory solvent forces become significant only at $z_{sep} < 2$ nm. Elastohydrodynamic simulations suggest that flat gold surfaces separated by 2.3 nm with 0.9-nm asperities should slide smoothly on a thin film of liquid lubricant molecules, albeit with nanometer-scale cavitated zones extending ~3 nm downstream persisting for ~0.1 nanosec at a sliding velocity of 10 m/sec. Solvation forces may also appear at separations below a few molecular diameters.

These results suggest that water-carrying nanocapillaries >2-4 nm in internal diameter should function largely in line with continuum models. Early theoretical calculations predicted that open-ended carbon nanotubes as small as 0.8 nm in diameter might act as "nanostraws" and could wick in molecules from vapour or fluid phases. Subsequent experiments confirmed that fluids with γ_{lg} <~ 190 x 10^{-3} N/m, including liquid sulfur, liquid selenium and nitric acid, are readily drawn into the inner cavity of carbon nanotubes with inside diameters of 48 nm through capillarity. This limit is sufficiently high to allow nanotube wetting by water (γ_{lg} ~ 72×10^{-3} N/m), most organic solvents ($\gamma_{lg} < 72\times10^{-3}$ N/m), and liquid acids (e.g., γ_{lg} ~ 43×10^{-3} N/m for HNO_3) which can then be used as low surface tension carriers to introduce dissolved solute into the nanotubes. Bulk nanotubes are readily wetted by water. Good wetting is favoured when the polarizability of the tube material is higher than that of the liquid. Pure metals like molten lead or mercury with γ_{lg} >~ 190×10^{-3} N/m do not wet carbon nanotubes and are not drawn in by capillarity. Such high-γ_{lg} liquids may be forced into the nanotubes by applying a hydrostatic pressure p_{force} given by the Laplace equation:

$$P_{force} \sim \frac{2\gamma_{lg}\cos(\theta)}{r_{cap}}$$

Thus, liquid mercury with $\gamma_{lg}=490\times10^{-3}$ N/m can be pushed into an r_{cap}=0.5 micron tube by applying p_{force} ~ 20 atm to the fluid; p_{force} ~ 2000 atm for r_{cap}=5 nm.

By 1998, detailed molecular dynamics simulation of fluid flow inside 1.3–1.6 nm diameter carbon nanotubes had been a subject of active research. One major difference between macroscale and nanoscale fluidic systems is that in nanomachines the tube walls may be considerably less rigid and may flex and resonate. Nanomachine designers must take into account the effects of these vibrations as they are transmitted through nanoscale structural elements attached to the tubes. Furthermore, tube wall motions are sometimes strongly size-dependent. The simulations confirmed that fluids flow faster through rigid tubes than through floppy tubes, as predicted by classical hydrodynamics theory. At high flow velocities we may expect nanotubes to exhibit buckling modes, variable patency, progressive wave propagation, sluicing, flow-limiting flutter, and stall, by analogy to the fluid mechanics of compliant tube flow in human veins. Carbon nanotubes may be rigidified by imposing a stretching tension, or possibly may be replaced by stiffer diamondoid tubes. If the fluid has more than one kind of atom, size-mass effects and differing interaction ranges or strengths may cause one or more species to segregate at the walls or to flow at different rates.

Material flow through nanotubes has been well-exploited by biology, including the bacterial sex pili, bacterial anal pores and cytoprocts, cellular excretory canals (e.g., in the glandular cells of the pancreas), 10-100 nm intranuclear nucleolar canals, the 20-40 nm wide hollow tubular fluid transport network that forms the smooth intracellular endoplasmic reticulum, the 75 nm diameter nutrient-transporting tubovesicular membrane network of the human malaria parasite, the >100-nm wide fluid-carrying human bone canaliculi, and prelymphatic tissue channels.

The syringelike T4 bacteriophage tail assembly is perhaps the best-studied example of biological nanotube flow. The 100-nm long, 20-nm wide cylindrical T4 tail assembly, made of 15 different proteins joined in 24 annular segments with an 8-nm inside bore, has a set of small fibres near the tip that attach to the plasma membrane of the host cell. After a lysozyme-like enzyme opens a breach in the host cell wall, the tail sheath thickens and contracts, inserting a ~2.8 megadalton hollow core protein nanotube (80 nm long, 7 nm wide, 2.5 nm inside bore) through the host cell integument. The protein nanotube is then uncorked in response to chemical signals, and a large-molecule (mostly putrescine and spermadine) pressure of ~30 atm ejects a single 70-micron long, 2-nm wide DNA thread (~50 Percent of head volume) through the 2.5-nm nanotube aperture and out into the cell typically in ~3 seconds, a mean flow velocity of ~23 microns/sec. The double-stranded DNA thread rotates ~4000 times around its axis as it emerges. Under optimum conditions, initial injection velocity may start as high as 360 micron/sec with a minimum injection time of 0.23 sec.

The tensile strength of liquids may also be important in nanorobot internal fluid flows, even in the complete absence of capillarity. Completely liquid-filled wetted tubes with no gas pockets make a continuous liquid column that can exhibit a large internal tensile strength $K_{liquid} \sim 4\ \gamma_{lg} / x_{molec}$, where $x_{molec} \sim 10$ nm is the approximate maximum range of intermolecular forces. Taking $\gamma_{lg} \sim 72 \times 10^{-3}$ N/m for water, $K_{liquid} \sim 300$ atm. There is experimental evidence that water saturated with air but denucleated by high pressures does exhibit a tensile strength on the order of ~300 atm. This tensile strength, and not capillarity, is credited for the ability of tall trees to pull sap up to 115 meters high, far exceeding the maximum 10.33 meters that water can be pulled vertically at the Earth's surface using a vacuum pump. M. Zimmermann reviews the sap transport system; the maximum pull recorded experimentally in trees is 120 atm.

Pipe Flow

Cellular and nuclear microinjection using fine glass pipette tips as small as 200 nm in diameter is commonplace in the experimental biological sciences. In small tubes or nanoinjectors with $r_{cap} >> 1$, nanoscale fluid flow behaviour is approximated reasonably well by the classical continuum equations. Continuum flow is governed by the famous Hagen-Poiseuille Law (or more commonly, Poişeuille's Law), derived from the Navier-Stokes equations, which states that a pressure difference of Δp between the ends of a tube of radius r_{tube} and length l_{tube} will move an incompressible fluid of absolute viscosity η in laminar flow at a volume rate of:

$$V_{HP} = \frac{\pi\, r_{tube}^4 \Delta p}{8 \eta\, l_{tube}} = \pi r_{tube}^2\, v_{flow}\ \left(m^3 / \sec\right)$$

The subsonic mean fluid velocity (v_{flow}), flow power dissipation (P_{flow}), and the flow time of an aliquot of fluid through the length of the entire tube (t_{flow}) then follow directly from Eqn. as:

$$V_{flow} = \frac{r_{tube}^{2}}{8\eta 1_{tube}} \left(m/sec\right)$$

$$P_{flow} = \frac{\pi r_{tube}^{4}}{8\eta 1_{tube}} \left(watts\right)$$

$$t_{flow} = \frac{8\ \eta\ 1_{tube}^{2}}{8_{tube}^{2}\Delta p} \left(sec\right)$$

Taking η=0.6915×10^{-3} kg/m-sec for pure water at T=310 K, a nanotube with r_{tube}=10 nm, l_{tube}=1 micron, and Δp=1 atm passes fluid at 'V_{HP}=0.6 micron^3/sec, v_{flow}=2 mm/sec, with t_{flow}=0.5 millisec. The incompressibility assumption generally holds in liquids and in gas flows where $\Delta p << p_{tube}$, where p_{tube} is the head pressure at the tube entrance.

The above equations describe the resistance to Poiseuille (laminar) flow in a pipe, which is the minimum of resistance of all possible flows in a pipe. If the flow becomes turbulent, the resistance increases. The determinative parameter is a dimensionless quantity called the Reynolds number, N_R, which is the ratio of the inertial pressure ($\sim \rho\, v_{flow}^2$) to the viscous pressure ($\sim \eta\, v_{flow} / r_{tube}$) in the flow of a fluid of density ρ, or:

$$N_R = \frac{\rho v_{flow} r_{tube}}{\eta}$$

A large Reynolds number implies a preponderant inertial effect and the onset of turbulence; a small Reynolds number implies a predominant shear effect and the maintenance of laminar flow. Reynolds found that the transition from laminar to turbulent flow typically occurred at $N_R \sim$ 2000-13,000, depending upon the smoothness of the entry conditions. The lowest value obtainable experimentally on a rough entrance appeared to be $N_R \sim$ 2000. However, when extreme care was taken to establish smooth entry conditions the transition could be delayed to Reynolds numbers as high as 40,000.

Even assuming the more conservative $N_R \sim 2000$ figure, it is clear that subsonic flow in nanoscale pipes will almost always be laminar. For a pipe with r_{tube}=1 micron conveying water at 310 K (r=993.4 kg/m^3), the onset of turbulent flow (taking $N_R \sim 2000$) occurs at v_{turb} > 1400 m/sec, very nearly the speed of sound in water. In gases or liquids of lower density, or in pipes of narrower bore, or if a less conservative transitional Reynolds number is available due to superior design, then v_{turb} grows still larger. Thus turbulence is primarily a high-velocity, large-tube phenomenon.

In such turbulent flow situations with $N_R >\sim 1000$, a well-known empirical formula for the volume flow rate is:

$$\dot{V}_{turb} \sim V_{HP} / Z \ \left(m^3 / sec\right)$$

where the turbulence factor Z=0.005 $N_R^{3/4}$. Thus for N_R=3000, $'V_{turb} \sim 0.5\ 'V_{HP}$. Nevertheless, even in turbulent conditions of the general flow, fluid motion nearest the tube wall remains laminar in a thin layer of thickness x_{lam}, often estimated as:

$$x_{lam} - \eta \left(\frac{501_{tube}}{\rho r_{tube} \Delta p} \right)^{1/2}$$

For flowing water at 310 K and r_{tube}=10 micron, l_{tube}=100 micron, and Δp=10 atm giving $v_{flow} \sim 200$ m/sec and $N_R \sim 3000$, then $x_{lam} \sim 0.5$ microns.

Poiseuille's Law assumes a rigid pipe, an assumption that may not hold for some nanotube designs and which certainly does not hold for human blood vessels, especially the veins. A complete treatment of fluid flow in elastic tubes. However, for steady laminar flow in an elastic tube with wall thickness h_{wall}, radius r_{tube}, length l_{tube}, p_0 and p_1 the pressures at the entry and exit ends of the tube, and E_{wall}=Young's modulus of a Hookean wall material, then the volume flow rate through the tube may be approximated by:

$$\dot{V}_{plastic} = \left(\frac{\pi c_h r_{tube}^4}{24 \eta^1_{tube}} \right) \left[\left(1 - \frac{P_1}{c_h} \right)^{-3} - \left(1 - \frac{P_0}{c_h} \right)^{-3} \right] \left(m^3 / \sec\right)$$

where $c_h = E\, h_{wall} / r_{tube}$. Taking $\eta = 0.6915 \times 10^{-3}$ kg/m-sec for water at 310 K, $h_{wall} = 2$ nm, $r_{tube} = 10$ nm, $l_{tube} = 1$ micron, $p_1 = (0.5\, p_0) = 1$ atm, and $E = 10^7$ N/m^2, then '$V_{elastic} = 0.3$ micron^3/sec. Care must be taken in applying this formula because many biological tube materials such as blood vessel walls are non-Hookean and exhibit a linear pressure-radius relationship instead. For small elastic deformations, the volume flow rate of such non-Hookean tubes may be approximated by:

$$\dot{V}_{hv} = \frac{\pi\, r_{tube}^4}{20\alpha\eta l_{tube}}$$

where α is the compliance constant, measured experimentally in feline pulmonary veins as 1.98–2.79 m^2/N for $r_{tube} = 50$–100 microns down to 0.57–0.79 m^2/N for $r_{tube} = 400$–600 microns. Poiseuille's Law also does not strictly hold for fluids seeded with polymers in concentrations as low as 10^{-4}–10^{-5} by weight, wherein drag may be reduced by a factor of 23.

It bears repeating here that as pipes get very small, they can clog more easily due to van der Waals adhesive forces, so it becomes increasingly important to engineer interior vessel surfaces for minimum adhesivity with respect to all materials likely to be transported through the pipes.

NANOMANIPULATORS

Numerous simple actuators have been developed for microelectromechanical systems or MEMS. A few of these designs might possibly be useful in early nanomedical systems. All major components of an articulated micromanipulator having multiple degrees of freedom, workspaces on the order of 1 mm^3, positional resolution up to ~10 nm, operating frequencies up to 10 KHz, and slewing speeds up to 4 mm/sec have been demonstrated experimentally; complete systems have even been proposed.

Actuation and manipulation may be very broadly defined. Self-assembling and self-disassembling ~300 micron gold foil cubes have been fabricated and cycled repeatedly at ~1 Hz; "silicon origami" has also been described. Nonmechanical actuation has also been explored. In the Magnetic Stereotaxis System, a helmet with a cubic array of six superconducting coils is used to apply a 0.2 N force to a 3.2-mm wide, 4.7-mm long permanently magnetized cylindrical pellet that is embedded in brain material, causing stepped movement of the pellet through the brain material to a specified location with ~1 mm placement accuracy without any direct physical contact.

However, in the usual definition of a manipulator, an ergomechanical transducer converts chemical, electrical, mechanical, acoustical, or some other form of energy into the mechanical energy of a manipulatory device which then exerts useful forces on objects or materials in the environment. In most cases a force-transmitting physical structure is required which may include both heavy load-bearing elements such as girders, struts, pneumatic tubes or rotary joints, and fine control elements such as stress cables, valves or switches. An end-effector is usually employed to focus and redirect the application of force at the distal end of the manipulation device, or to perform specialized tasks. Sensors permit feedback control of manipulator motions as well as verification that the assigned manipulatory task has been properly executed. Large numbers of manipulators can form ciliary arrays, allowing convenient mass transport of materials; other classes of manipulators may be useful in bulk disassembly of materials.

The diverse concepts outlined in this Section demonstrate the enormous range of possible nanomanipulator designs. These concepts are not intended as specific engineering proposals but rather as illustrations of certain useful classes of manipulators representing alternative design pathways that could provide the desired capability. No effort has been made to produce optimal systems that minimize energy dissipation, maximize speed, minimize parts count, or simplify manufacture.

Biological Cilia

Biological cilia are motile, hairlike cylindrical processes typically 200–300 nm in diameter, 2–20 microns in length, and

$\sim10^{-16}$-10^{-15} kg in mass. They are found in large numbers on some cell surfaces. Unicellular organisms use cilia both for locomotion and for food collection, while multicellular organisms use cilia primarily for mass transport of the environment past the cell.

Each cilium is bounded by an extension of the cellular plasma membrane, rising from a cytoskeletally-anchored centriole-like basal body consisting of an array of 9 triplets of common-walled microtubules ~150 nm in diameter and 300-500 nm in length. Sprouting upward from the basal body is the primary ciliary structure, called the axoneme. The axoneme is a cylinder of microtubules with 9 outer doublets of microtubules and two additional complete microtubules in the centre, often called the central pair. The 9 outer doublets are extensions of two of the three microtubules comprising each of the nine basal body triplets. Each outer doublet of the axoneme consists of the A tubule, the B tubule, and a common wall containing the protein tektin.

A set of 1-2 megadalton dynein side arms project outward from each of the A tubules, reaching clockwise toward the B tubules of the adjacent doublet. The dynein arms occur in pairs, one inner arm and one outer arm, spaced at regular ~30 nm intervals along the microtubule. Nexin protein interdoublet links at wider intervals join the outer doublets making circumferential rings, and radial spokes project inward from each of the 9 microtubule doublets. Microtubules do not change length during ciliary movement. Rather, the stalk of each dynein arm attaches to and detaches from the adjacent B tubule in an ATP-driven cyclic process, pulling the stalk the length of one dimer along the tubule with a velocity of ~14 microns/sec. The maximum force generated by a dynein arm has been measured as ~1 pN. The differential stress of adjacent doublet columns causes the axoneme to flex. Selective sidearm bonding and unbonding allows the cilium to generate a bend anywhere along its length.

In early experiments, bullfrog gullet ciliary surfaces developed a mechanical motive power of ~0.01 watts/m^2;

assuming 1-10 micron^2/cilium across the gullet surface gives an estimated mechanical power output P_{cilium} ~ 0.01-0.1 pW/cilium. Similarly, hair cell cilia of the human inner ear respond to a minimum 2×10^{-5} N/m^2 (2×10^{-10} atm) pressure with a minimum 0.1 nm deflection at an applied force of ~10^{-3} pN, indicating a mechanical stiffness of ~10^{-5} N/m. Assuming a maximum D_p ~ 10^5 watts/m^3 power density in nonmuscular working tissue and a ciliary volume of ~0.1-1 micron^3 also gives an estimated mechanical power output of P_{cilium} ~ 0.01-0.1 pW/cilium.

A simple cilium 12 microns long may beat at a frequency n_{cilium} ~30 Hz with an angular velocity of 12 deg/millisec and a tip velocity of 2500 microns/sec in the effective stroke, while the bend in the recovery stroke propagates at ~350 microns/sec.

IN VIVO LOCOMOTION

One of the most important basic capabilities a medical nanorobot may possess is the ability to move about inside the human body. At its most simple, this movement may be purely statistical, with nanodevices carried along with the natural ebb and flow of bodily fluids. At the other extreme, nanorobot locomotion may be highly deterministic, including powered drive mechanisms, mapping and active navigation, and traverses of diverse histological territories having markedly different mechanical and chemical characteristics. The subject matter is huge and quite impossible to cover fully in a single Chapter. As a result, the discussion here is merely a preliminary survey of the most important issues and challenges of in vivo locomotion, suggesting promising new areas for future research.

Rheology of Nanorobot-Rich Biofluids

The study of biofluid flow, or biorheology, is useful in nanomedicine because it is necessary to understand the flow characteristics of fluids through which nanorobots must navigate – whether a nanorobot is actively swimming or is simply drifting with the flow, or whether the nanorobot is traveling in an ordinary Newtonian fluid such as water or in more complex viscoelastic biofluids such as mucus or saliva. Additionally, the presence of large numbers of nanorobots may dramatically alter bloodstream viscosity, with important medical consequences. The following discussion examines biofluid and whole-blood viscosities, the radial distribution of blood elements in blood vessels, viscosity and bloodstream velocity profiles of nanorobot-rich blood, and hematocrit reduction in narrow blood vessels.

BIOFLUID VISCOSITY

Viscosity is a measure of the resistance of a fluid to shearing when the fluid is in motion. Consider a plate of surface area A moving parallel to a fixed plane surface with constant velocity v, being pushed laterally by a constant force F. A fluid of viscosity h fills the volume between the two surfaces, which are separated by a distance d. The fluid layer nearest the moving plate also moves at velocity v and the layer nearest the stationary plane remains stationary. In between the two surfaces, the velocity increases linearly with distance, establishing a constant gradient. This gradient is usually called the shear rate, or 'γ=v/d (m/sec–m, or sec^{-1}). The absolute viscosity may then be defined by:

$$F/A = \eta \dot{\gamma} \quad \left(N/m^2\right)$$

where η has MKS units of N-sec/m^2, Pascal-sec, or, more simply, kg/m-sec. In a Newtonian fluid, the shear stress F/A (N/m^2) increases linearly with shear rate 'γ, so that viscosity η is constant over a wide range of shear rates. Many common fluids such as air, water, saline and blood serum closely approximate the ideal Newtonian fluid. The viscosity of solutions of molecules is related to the diffusion coefficient.

The viscosity of an ideal gas is independent of density and independent of pressure between 0.01-10 atm; at higher pressures, intermolecular interactions lead to higher viscosities. Theory predicts that gas viscosity ~ $T^{1/2}$, but a somewhat larger temperature exponent is obtained experimentally for real gases. In contrast, the viscosity of liquids increases with rising pressure and decreases with rising temperature. In normal liquids the temperature dependence is approximated by Andrade's formula which gives:

$$\eta \sim k_v \, e^{Ev/kT}$$

where the activation energy for viscosity E_v~25 zJ and the constant k_v~2.1×10^{-6} kg/m-sec for pure water at 1 atm. Nonelectrolytes dissolved in water generally cause viscosity to rise, while solvation of electrolytes may increase or decrease viscosity. Large asymmetric solute molecules increase viscosity more than an equal mass of small spherical molecules. The threshold between solid and liquid is generally taken as η~10^{14} kg/m–sec.

Most biofluids are viscoelastic and non-Newtonian, with apparent viscosity η_a varying with shear rate and displaying other nonlinear characteristics such as hysteresis, relaxation, and creep. Saliva behaves more like an elastic body than like water. Because of its high molecular weight, DNA solution is viscoelastic even at low concentrations. Sex glands produce viscoelastic fluids, including semen and uterine cervical mucus (η_a varies ~20 Percent during the menstrual cycle). Human synovial fluid becomes increasingly incompressible at higher pressures, with η_a~10 kg/m-sec at 'γ~0.1 sec^{-1} declining to η_a~0.1 kg/m-sec at 'γ ~ 10 sec^{-1}, and to η_a~0.001 kg/m-sec at 'γ ~ 10,000 sec^{-1}. Viscoelasticity is also an important property of respiratory tract mucus, which typically shows an η_a~1 kg/m–sec at low shear rates near 'γ~0.1–1 sec^{-1}, falling to η_a~0.01 kg/m-sec at high shear rates near 'γ ~ 100-1000 sec^{-1}. Even ice is a viscoelastic material, with η_a ~ 10^{10}-10^{13} kg/m–sec for 'γ ~ 10^{-3}–10^{-7} sec^{-1} at 262 K; ice viscosity varies with temperature as described by Andrade's formula, with E_v ~ 110 zJ as determined experimentally for pure ice, and taking k_v~6.3×10^{-4} kg/m-sec at high shear rate.

Cytoplasm is a complex viscoelastic material having a continuous liquid phase plus various suspended particles, granules, and membranous structures. More precisely, the cytomatrix is a mixed-phase body composed of a fibrillar network penetrated by a solution. As a result, viscosity is different in the various phases. From flow behaviour, the viscosity of *E. coli* protoplasm was estimated as ~1000 kg/m-sec; from measured diffusion rates of sucrose, dextran, and β-galactosidase, the apparent viscosity was 3-4$\times 10^{-3}$ kg/m-sec. Cytoplasm is inhomogeneous and anisotropic at many levels of organization.

Viscosity of Whole Blood

Human blood is a suspension of cells, in plasma. The plasma is ~90 Percent water by weight, 7 Percent plasma proteins, 1 Percent inorganic materials, and 1 Percent other organic substances. The cellular component is essentially all erythrocytes, or red blood cells, with white cells of various categories comprising less than 1/600th of the total volume of cells and platelets less than 1/800th.

Pure blood plasma is a Newtonian viscous fluid, with $\eta_{plasma} = 1.1 \times 10^{-3}$ kg/m–sec at 310 K. Blood plasma is the environment normally encountered by micron-sized bloodstream-traversing nanorobots as they move between blood cells.

On the other hand, "whole" blood is the complete natural mixture of blood plasma and blood cells. In blood vessels whose diameters are much larger than the size of blood cells, or in the case of mechanical systems with dimensions much larger than the characteristic sizes of blood cells, whole blood must be treated as an homogenous non-Newtonian fluid. The bulk viscosity of whole blood decreases with rising shear rate and increases with rising hematocrit (Hct), the percentage of blood volume occupied by red cells. For large blood vessels, the effect of hematocrit on viscosity for Hct <~ 45 Percent has been modeled experimentally by Cokelet et al as:

$$\eta_a / \eta_{plasma} = (1 - Hct)^{-2.5}$$

with Hct expressed as a fraction. This formula, which also gives good results for RBC–sized oil/water emulsions in the same volume fraction range, is only useful up to volume fraction of ~10 Percent in the case of rigid-particle suspensions, whose viscosity behaves markedly differently from that of RBC suspensions.

It has long been known that human red blood cells can form aggregates known as rouleaux, in which the discoid RBCs adhere loosely in a "stack of coins" configuration at shear rates <100 sec^{-1}. Formation of linear and branched chain aggregates depends on the presence of the cell-surface cross-linking proteins fibrinogen and globulin in the plasma. The lower the shear rate, the more prevalent are the aggregates. As the shear rate goes to zero, it is speculated that human blood becomes one big aggregate, which then may behave as a viscoelastic or viscoplastic solid.

As shear rate increases, the rouleaux tend to break up. Individual red cells also deform slightly, elongating and lining up with the streamlines. These two effects combine to reduce blood viscosity with rising shear, which shows the relative viscosity η_a/η_{plasma} for human blood at Hct=45 Percent as a function of shear rate. Note that for rigid particles, bulk viscosity is essentially independent of shear rate.

Normal blood vessel wall shear rates in physiological bloodflow range from 50-700 sec^{-1} in the larger arteries to 250–2000 sec^{-1} in the smallest arteries and capillaries, and 20–200 sec^{-1} in large and small veins. At shear rates >100 sec^{-1} where RBC aggregation ceases to be important, pure whole blood remains fluid even up to 98 Percent RBC concentration by volume.

Radial Distribution of Blood Elements

At low shear rates, red cells aggregate into rouleaux and migrate inward, forming a network of aggregates in the core of the tube. Individual rouleaux may incorporate 10–20 red cells, or more, creating by far the largest cellular elements normally present in the blood. At the highest shear rates, the rouleaux break up entirely into single red cells, and the red cells then distribute themselves more uniformly in the radial direction.

The radial distribution of white cells is also a function of flow conditions. At low shear rates, under conditions allowing red cell aggregation, the white cells are displaced to the periphery of the flow by the much larger red cell rouleaux. At the highest shear rates, white cell concentration is highest along the tube axis, displacing some red cells, since most white cells are larger than individual red cells.

The radial distribution of the local platelet concentration during blood flow in tubes has also been investigated. In plasma containing only platelets, the platelet distribution is radially uniform. However, in whole blood flow under all shear conditions, the platelet concentration is highest near the vessel wall. In arterioles, the platelet number density is about two times higher near the wall than in the centre of the vessel. Platelets are much smaller than either red cells or white cells, thus tend to be crowded out of the centre whenever red cells or white cells are present.

These experimental observations are consistent with the general principle that during blood flow in a vessel, the largest particles, or "flow units," move toward the axial region, leaving the smallest particles more concentrated at the periphery. In most cases, bloodborne medical nanorobots will be ~2 microns in diameter or smaller. As such, they will normally constitute the smallest particles in the bloodstream. More than any other blood element, free-floating nanorobots should tend to migrate nearest the blood vessel walls, although at high shear rates the radial diffusivity is significantly increased due to local fluid motions generated by red cell rotations. Small molecules and complexes such as lipoproteins are less subject to erythrocyte-induced diffusivity enhancement, hence may migrate toward the walls.

Viscosity of Nanorobot-Rich Blood

The presence of large numbers of relatively rigid nanorobots dramatically alters bloodstream viscosity. The relative viscosity η_a/η_{plasma} for human blood at 298 K with shear rate >100 sec^{-1} as a function of particle volume fraction, compared to the relative viscosity of suspensions of latex rigid spheres, rigid disks, emulsion droplets, and sickled erythrocytes, as determined experimentally. A 50 Percent suspension of micron-sized rigid nanorobots will increase blood viscosity by a factor of ~350, seriously impeding flow especially in the smaller vessels. However, a plasma suspension of microspheres at a ~10 Percent particle volume fraction has a relative viscosity indistinguishable from Hct=10 Percent whole blood. This suggests a conservative 10 Percent volume-fraction limit for the maximum bloodstream concentration of medical nanorobots, a limit that may also ensure free flow of the fluid.

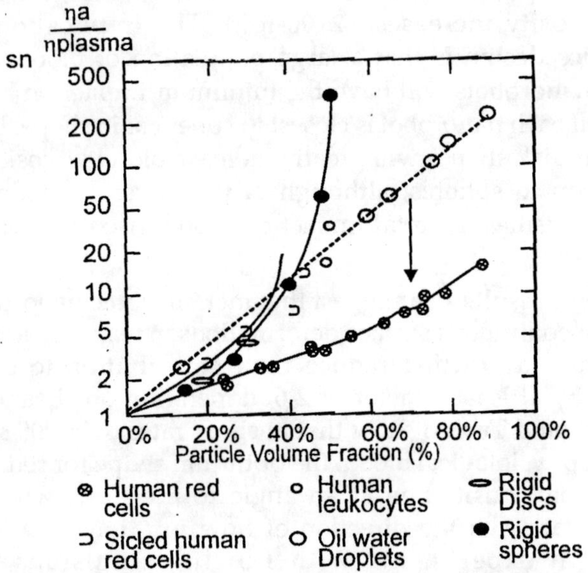

Fig. Viscosity of nanorobot-rich human blood at high shear rate

Relative viscosity also depends on particle size, though the effect due to the presence of medical nanorobots is usually minor. As a conservative upper limit in the smallest vessels:

$$\eta_a / \eta_{plasma} = \left[1 - \left(\frac{D_{namo}}{d_{tube}}\right)^4\right]^{-1}$$

where D_{nano} (=2 R_{nano}) is the maximum nanorobot diameter (radius) and d_{tube} (= 2 r_{tube}) is the blood vessel diameter. Taking d_{tube}=8 microns, then η_a/η_{plasma}=1.0002 for D_{nano}=1 micron, or 1.07 for D_{nano}=4 microns the largest bloodborne nanorobot. In larger blood vessels, this effect is even smaller for micron-sized medical nanorobots.

Relative blood viscosity also depends on nanorobot shape. Chien's measurements of effective viscosity as a function of particle shape in dilute suspensions found that minimum viscosity is achieved by hard spheres or by 1:1 hard cylinders. Thin disks or long cylinders have higher viscosity. 10:1 rods produce a suspension with ~10 times higher viscosity than a suspension containing an equal volume of spheres; for 100:1 rods, viscosity increases ~2500-fold. The implication for nanodevice design is that a large population of bloodborne medical nanorobots will have the minimum impact on blood viscosity if each nanorobot is closest to spherical in shape. Long rod or flat disk shapes will greatly increase blood viscosity, in comparison to spheres, although at the lowest nanodevice number densities the total impact on blood viscosity may be negligible.

Chien's results also suggest that metamorphic nanorobots capable of continuous surface deformations in response to flow conditions may further reduce their contribution to blood viscosity by at least a factor of 2-6, depending on shear rate. Goldsmith and Turitto show that at shear rates over 200 sec^{-1}, typical in physiological blood, the optimum shape for red cells is ellipsoidal, positioned at an angle to the flow, with the surface rotating in the direction of flow in a tank-tread-like motion. In experiments with flowing emulsions, the deformation of a liquid droplet results in its migration across the streamlines away from the tube wall. Thus in physiological blood over the whole range of normal hematocrits and typical flow rates, there is a plasma-rich or "plasmatic" zone d_{plasma} ~ 2–4 microns deep at the walls of vessels whose diameters exceed 100 microns.

Such lateral migration is not observed with small rigid particles of any shape at high concentrations and at low Reynolds numbers N_R <~ 10^3. However, for vessels with N_R >~ 1, inertial effects do come into play and rigid free-floating nanorobots will be pushed away from the wall to produce a particle-free zone. The thickness of this "plasmatic" zone d_{nano}

decreases sharply with increasing nanorobot concentration (Nct). At Nct=2 Percent, $\delta_{nano} \sim 0.3\ r_{tube}$; at Nct=10 Percent, $\delta_{nano} \sim 0.1\ r_{tube}$; at Nct=30 Percent, $\delta_{nano} \sim 0.01\ r_{tube}$.

In terms of individual nanorobot motion, a rigid sphere initially placed near the tube wall migrates inward, while a rigid sphere placed near the tube axis migrates outward. Known as the "tubular pinch effect," rigid spheres started in either position converge to an intermediate equilibrium radius position of $r_{eq} \sim (0.6\text{-}0.7)\ r_{tube}$ for $R_{nano}/r_{tube} << 1$, or $r_{eq} \sim 0.5\ r_{tube}$ (farther from the wall) for $R_{nano}/r_{tube} \sim 0.25$. By analogy with Brownian translational diffusion, a radial dispersion coefficient D_r may be defined as $\Delta r = (2\,\tau\,D_r)^{1/2}$ (meters), where Dr is the RMS radial displacement of a bloodborne object in an observation time τ. The analogy is imperfect because these radial motions are not random, but are due to multibody collisions determined by the local velocity gradient, particle concentration, and surface deformations of the objects. At any given concentration, displacements are greatest at radial distances between 0.5–0.8 r_{tube}. For local shear rates of 5–20 sec^{-1} and volume concentrations from 20 Percent–70 Percent, $D_r = 1\text{-}20 \times 10^{-12}\ m^2/sec$ both for red cells and for rigid 2-micron diameter microspheres, and $3\text{-}86 \times 10^{-11}\ m^2/sec$ for platelets in whole blood, as determined experimentally. Thus the mean time for a nanorobot (R_{nano}=1 micron) to migrate a radial distance $\Delta r \sim 1$ micron is $\tau \sim 25\text{–}500$ millisec - about an order of magnitude faster than simple Brownian diffusion.

Hematocrit Reduction in Narrow Vessels

Fahraeus found that when blood of a constant hematocrit Hct is allowed to flow from a large feed reservoir into a small tube, hematocrit in the tube (Hct_{tube}) decreases as the tube diameter decreases; Fahraeus and Lindqvist found a decrease in apparent viscosity when tube diameter is reduced to below 300 microns. Barbee and Cokelet crudely approximated the experimental data using the following empirical formula:

$$Hct_{tube} \sim \left(a\,I\left(d_{tube}\right) + b\right) Hbt\,(\%)$$

where a=0.196, b=-0.117, and d_{tube} is expressed in microns. Given a normal human male hematocrit of Hct=46 Percent, in smaller vessels this falls to Hct_{tube}=36 Percent at d_{tube}=100 microns, 25 Percent at 30 microns, and 13 Percent at 8 microns. Surveying the literature, Gaehtgens concludes that minimum hematocrit Hct_{tube} generally occurs at d_{tube}~15-20 microns, which Cokelet suggests marks the transition from multi-file flow to single-file flow among the red cells. Gaehtgens also showed that the relative viscosity of human RBC suspensions reaches a minimum at about 5-7 microns.

Hematocrit decreases in small blood vessels for several reasons. First, a cell-free layer approximately equal to RBC radius exists near the wall, so the smaller the vessel, the larger the fraction of volume occupied by this layer, hence the lower the hematocrit. Second, a vessel side branch that draws mainly from the cell-free layer produces a lower hematocrit in that side branch. Third, red cells are elongated and oriented along the direction of shear flow, making it less likely that they will enter a side branch aligned perpendicular to the direction of flow. All three factors should be less important for near-spherical rigid nanorobots that are smaller than RBCs, hence any reduction in nanocrit during passage through narrow blood vessels should be quite modest.

The velocity of a cell or nanorobot located on the axis of a small vessel is greater than the mean velocity of the suspending fluid, but is always slightly less than the fluid in the immediate vicinity on the axis. The simplest model is the stacked-coins model discussed by Whitmore, wherein the nanorobot velocity v_{nano} is given by:

$$\frac{v_{nano}}{v_{flow}} = 2\left[1 + \left(\frac{D_{nano}}{d_{tube}}\right)^2\right]^{-1}$$

Taking d_{tube}=8 microns and v_{flow}=1 mm/sec for a typical capillary, a nanorobot with diameter D_{nano}=2 microns has v_{nano}=1.88 mm/sec, which is faster than v_{flow} but is slower than v_{max}=2 mm/sec.

IN CYTO LOCOMOTION

Once inside a cell, a motile nanorobot must navigate a cluttered and highly viscous cytomatrix-rich environment.

Any of the methods described for cytoambulation, histonatation, or cytopenetration may also be employed in modified form within the cell. The Reynolds number of an L~1 micron object inside a red blood cell is $N_R \sim 10^{-6}$. For a nanorobot inside a free leukocyte, $N_R \sim 10^{-9}$; inside an *E. coli* bacterium, $N_R \sim 10^{-11}$, taking the higher viscosity into account.

There are many natural transport mechanisms inside cytomatrix-rich cells which may serve as analogs for in cyto nanorobot locomotion. Vesicles and granules ~100 nm in diameter or larger are carried at a peak speed of up to ~2 microns/sec (although mean unloaded kinesin motor speed is usually 0.5-0.8 microns/sec) on the back of a 60-nm kinesin transport molecule that takes 8-nm ATP-powered steps along microtubule tracks running throughout the cell with a stall force of 5-7 pN; typically kinesin takes ~100 steps along a microtubule, then lets go. Nanorobots could be designed to brachiate along these tracks as well. Inside giant amoebas, mitochondria measuring 1-3 microns in length are carried along microtubules at speeds up to 10 microns/sec, and the pseudopods of fibroblasts can also extend at ~10 microns/sec. Faster speeds may be possible for well-designed nanorobots, but an upper limit of ~10 microns/sec through fluid-rich intracellular clear channels seems reasonably conservative.

It is well-known that several pathogenic bacterial species, once free in the cytoplasm of a human cell, propel themselves through the cytosol using a continuous actin polymerization process that takes place at one pole of the bacterium. Actin assembly is visible as a tail of polymerized F-actin that remains stationary in the cytosol while the bacterium moves ahead. The polymerizing tail rectifies the random thermal motions of the bacterium, preventing it from diffusing backwards while permitting forward diffusion; thus the tail doesn't actually "push" the bacterium forward. Actin-based motility is mediated by a single bacterial protein — ActA in *Listeria* and IcsA/VirG (120,000 daltons) in *Shigella* — localized in the polar regions of the bacterium. Actin microfilaments (and tubulin) typically self-assemble at ~0.11 micron/sec; actin polymerization gives a stall force of ~10 pN per fibre, sufficient to drive a 1-micron object at ~1.5 micron/sec against a ~1 pN load in free fluid cytoplasm. This probably defines the top speed for actin-based bacterial mobility.

To progress through dense cytomatrical regions, motile nanorobots of similar size to bacteria must cut or detach cross-bridged cytoskeletal elements lying across the path ahead, then attempt to reattach or reconstruct those elements after the nanorobot has passed through the breach, because these elements lack sufficient elasticity to be pushed completely out of the way. Such cytoskeletal elements will most commonly include intermediate filaments and actin-based microfilaments. This procedure is crudely analogous to the process employed by fibroblasts transiting the ECM during wound repair. Fibroblast movement into cross-linked fibrin blood clots or tightly woven ECM requires an active proteolytic system that cleaves a pathway for migration; known enzymes serving this purpose include plasminogen activator, interstitial collagenase (MMP-1), the 72 kilodalton gelatinase A (MMP-2), and stromelysin (MMP-3).

Power requirements for intracellular mobility are typically modest but vary widely, depending mainly upon velocity, the type of cell, and the path chosen. In highly fluidic clear channels, viscosity should more closely resemble that of the red cell interior (h ~ 10^{-2} kg/m-sec), taking R_{nano}=1 micron and v_{nano} ~ 10 microns/sec, then P_{nano}/e Percent ~ (0.00002 pW)/e Percent for pure cytonatation. Nanorobots traversing pathways through filament-rich regions of the cell will experience much higher effective viscosities, on the order of h ~ 10-1000 kg/m-sec, but will also travel more slowly (e.g., v_{nano} ~ 1 micron/sec), giving a viscous resistance power requirement for the nanorobot body of P_{nano}/e Percent ~ (0.0002-0.02 pW)/ e Percent. A telescoping manipulator arm measuring 500 nm in length, performing the filament cuts and joins at n_{cut} ~ n_{join} ~ 400 Hz with ~2.5 microns of tool-tip travel per cycle, moves at a tip speed of ~1 mm/sec, producing ~0.025 pW of mechanical losses under no-load conditions; there are at least two operational arms, one fore and one (or more) aft. Multiple arms may be needed in a dense filament network. Another ~0.025 pW is required to overcome the viscous resistance against each moving arm, giving a total power cost of 0.1-0.3 pW for transfilamentary intracellular locomotion at v_{nano} ~ 1 micron/sec assuming locomotive efficiency e Percent=0.10.

Some of the difficulties and limitations of in nucleo locomotion have already been mentioned. The most important constraint is to observe a maximum speed limit that avoids mechanical chromatin damage. Although a 500 kilodalton protein would diffuse from one side of a 5-8 micron nucleus to the other in only ~5-6 sec, normal chromosomal movements during interphase have been estimated by observing the motions of individual centromeres, typically 0.002–0.003 microns/sec, and peak chromosome transport speeds during mitosis are ~0.1 micron/sec. A strict in nucleo speed limit of ~0.1 micron/sec on all nanorobot bodies and most exposed appendages appears prudent. A force limit of ~50 pN for intranuclear locomotion should also be observed — a 1-micron nanorobot picking its way through a filament-rich medium of net viscosity ~10 kg/m-sec requires the application of ~20 pN to overcome viscous drag at a travel speed of ~0.1 micron/sec.

Medical nanorobots will often be called upon to travel from place to place within the human body by actively swimming through the bloodstream, a process called sanguinatation. This Section describes the general nature of the process in terms of the Reynolds number, rotations and collisions with vessel walls and cellular blood components likely to be experienced by free-floating or powered nanorobots, nanorobot hydrodynamics, general force and power requirements for submersive swimming, various specific natation mechanisms, and some additional considerations regarding sanguinatation.

EX VIVO LOCOMOTION

Dental Walking

The exposed surfaces of natural dentition consist of a ~500 micron thick layer of enamel coating the exposed crown of each tooth inside the mouth. Enamel is a composite structure containing ~96 Percent inorganic material similar to hydroxyapatite, a dense bone mineral, embedded in an organic matrix of glycoprotein and a keratin-like protein. The mineral forms crystallites oriented into hexagonal cylinders with a horseshoe-shaped cross-section wrapping around the dentin, forming enamel prisms ~5 microns in diameter stacked in parallel congeries. These prismatic columns are directed vertically at the summit of the crown and horizontally along the sides of the tooth, pursuing a generally wavy course. There are no nutritive channels permeating the enamel and leading into the dentinal layer below, although dentinal tubules do extend from the pulp chamber into the dentin which may get extremely close to the occlusal surface of the tooth, and may be exposed by receding gums. However, ~0.4-micron-wide fissures separating enamel prisms are filled with an interprismatic substance of much higher organic content. The entire enamel surface is pockmarked by numerous micropores of irregular but generally oval shape, ranging from 0.1-2.0 microns in diameter and spaced ~1-2 microns apart. Thus a dental-walking micron-size nanorobot encounters a spongelike topography with holes and fissures comparable in size to the nanorobot diameter, and with gently rolling hills ~2 device diameters in height and ~5 diameters across.

Nanorobots may traverse tooth enamel using ambulatory techniques previously described in other contexts, including legged walking, rolling, or amoeboid locomotion using footpads for anchorage. Teeth may be clenched with a force up to 300-500 N, although forces of ~50 N are more common during normal mastication of soft foods. Assuming ~1 cm^2 of dental contact surface, shear forces at that surface during chewing or clenching are 0.5-5 x 10^6 N/m^2, close to the limit for noncovalent anchorage. Nanorobots seeking further refuge from shear forces may congregate in the valleys between enamel prisms or can seat themselves in the irregular micropores. E. Reifman notes that ~90 Percent of U.S. adults have multiple occlusal surface restorations composed of many different materials such as resin composites of various hardness coefficients, porcelain or gold crowns, amalgam fillings, composites, and the relatively newer but very popular glass polymer crowns, representing a substantial percentage of all exposed dental surfaces with which oral nanodevices must cope; yield strength of these materials at 0.1 Percent strain deformation typically ranges from 110-10,000 atm.

Are the forces of dental grinding sufficient to crush medical nanodevices? Tests of sized jeweler's grinding powders by the author confirm that even irregularly-shaped diamondoid particles ~3 microns and smaller apparently roll smoothly out of the way when ground between the teeth, whereas particles larger than ~3 microns cannot roll sufficiently and retain a sensible grittiness. In the case of a trapped nanorobot-size diamond block slowly being squeezed between opposed enamel surfaces, the Young's moduli for enamel and diamond are 7.5×10^{10} N/m^2 and 1.05×10^{12} N/m^2, respectively, so the enamel deforms ~14 times more than the diamond. Crystalline diamond can deform at least ~5 Percent before fracturing, diamond composites much more; a 50 Percent strain in each enamel surface produces a 3.6 Percent deformation across the diameter of the block, slightly below the conservative tolerance limit for diamondoid fracture strain. In the case of a 50-gram jaw moving at the maximum 0.1 m/sec clenching velocity that is suddenly brought to a halt via enamel deformation to a depth equal to a ~1 micron nanorobot radius across a 1 cm^2 enamel contact area, the jaw decelerates at ~1000 g's producing a kinetic force of ~500 N and a contact force of 5×10^6 N/m^2 (~50 atm), well below the ~10^{11} N/m^2 failure strength of solid diamond.

However, nanorobot crushing strength is an explicit design parameter which is normally significantly lower than for solid diamond if the device has large unbraced internal voids or diameters that are weak in compression. Such nanorobots may not be regarded as solid diamond structures as assumed above. The Euler buckling force for a solid cylindrical rod of outer radius R is proportional to R^4, but for a hollow cylinder with inner radius r the buckling force is proportional instead to (R^4 - r^4) during axial compressions.

Hence a trapped, thin-walled cylindrical nanorobot of radius R with r/R=0.90 buckles at only ~34 Percent of the force needed to buckle a solid rod of radius R, and thus could probably be crushed by static compression of the teeth. The oral cavity appears to be the only place in the human body where thin-walled medical nanorobots might plausibly be destroyed by mechanical grinding. Such destruction is avoided by nanorobots with composite diamondoid shells >~10 Percent R thick at a maximum allowable strain <~10 Percent, or by nanodevices possessing maps of mastication contact spots that actively avoid those spots during nanorobotic perambulations.

Epidermal Locomotion

The epidermis is a modified cellular surface, so most of the techniques described for cytoambulation are applicable to epidermal locomotion as well. There are so many special situations and obstacles to motility on skin that we can only mention a few of them here:

1. *Flaky Corneum* — Desiccated stratified epidermal cells crack, chip, or flake off entirely when rubbed or jostled with too much force. Nanorobots traversing a dead cell could find themselves flaked off into the air or onto the floor, especially if the surface is rubbed by the patient.
2. *Steep Topography* — Many physical features of the skin are huge in scale compared to a micron-size nanorobot and thus must be carefully circumnavigated. Dermal fingerprint ridges are ~500 microns wide and 20-50 microns deep. Hair follicle shafts are 50-100 microns in diameter. Scent-secreting apocrine glands are ~200 microns in diameter; eccrine glands are only 20 microns wide. Skin-dwelling fungi grow into irregular globs 5-20 microns in size. The tongue surface includes taste buds ~150 microns wide and tiny papillae measuring 30 microns wide and 80 microns long. At the micron scale, the topmost layer of the epidermis is littered with obstacles and protrusions, with electron micrographs showing a

surface that closely resembles a flaky puff pastry in appearance.

3. *Constant Motion* — The skin is almost always in motion at size scales up to ~1000 microns, folding and unfolding, stretching and tightening, twisting and curving, with normally distant surfaces being brought into and out of contact as the patient moves about normally.
4. *Dust Mites and Bacteria* — In eyelashes and eyebrows, the hairs of the upper lip, in the ears, and elsewhere on the body, dust mites measuring 150 microns x 300 microns survive by chewing up dead skin cells. Epidermal-walking nanorobots must be able to identify mite surfaces, and avoid climbing aboard them by mistake. Everyone inhales a few mites now and then; mite excrement is concentrated into fecal packets so small and light that they float in the air and may settle back onto the skin. Many other surface-dwelling ectoparasites and microfauna similarly must be avoided. Bacterial density over most of the body surface is quite low, $\sim 10^3/cm^2$; the count ranges from $\sim 10^2/cm^2$ on the palms and dorsa of the hands, up to 10^4-$10^5/cm^2$ on the hairy axilla, scalp, perineal regions, and beneath the distal end of the nail plate.
5. *Obstructions* — The skin may be covered with dirt, or grease, or cooking oils and fats, or sebaceous gland oils, which may be many tens or even hundreds of microns thick. Another potential obstacle is the sweat wash — the skin can pour out up to 2 liters/hour of perspiration. Nanorobots can easily crawl underneath the tightest-fitting clothing or rubber gloves, although a patch of superglue stuck to the skin could provide a formidable obstacle requiring overpassage.
6. *Itching/Crawling Sensations* — Tickling sensations attributable to isolated nanorobots traversing the skin are unlikely. Skin-crawling ~2 mm ants are readily

detected; ~100-micron mites are not. Absolute epidermal pressure stimulus thresholds, as measured by a laboratory esthesiometer, range from a low of 2000 N/m^2 at the tongue- and finger-tip, up to 12,000 N/m^2 on the back of the hand, 26,000 N/m^2 on abdomen, 48,000 N/m^2 on the loin, and a high of 250,000 N/m^2 on the thickest part of the sole of the foot. By comparison, the weight of a (100 micron)^3 nanorobot, if distributed across a (100 micron)^2 contact surface, is only ~1 N/m^2, quite undetectable on the skin. At a velocity of 1 cm/sec, the inertial force required to propel a 100-micron nanorobot is ~1 nN. This additional force, if distributed over 10 footpads each of area 100 micron^2, gives a shear pressure of ~1 N/m^2 across all contact surfaces, also undetectable; the sensible threshold velocity for such a nanorobot is ~45 cm/sec. Skin sensor frequency response is << KHz, whereas skin walkers <~10 microns in size may employ n_{leg} >~ KHz leg motions; the weight of a 1 micron^3 nanorobot produces only ~0.01 N/m^2 of contact pressure.

Another way to avoid obstacles is to hop over them in the manner of a jumping flea, an insect that subjects itself to an acceleration of ~200 g's while leaping ~130 times its own dimension. Diamondoid saltators could tolerate much higher accelerations, although it is important not to disturb the takeoff surface; also, airborne ex vivo trajectories may be only poorly controlled.

What about dislodgement forces? Assuming a headwind of v ~ 1 m/sec and a d ~ 10 micron boundary layer, then the shear stress on a 100-micron wide nanorobot is F/A ~ 100 N/m^2 (1000 nN) in water, 2 N/m^2 (20 nN) in air. A single 10 micron^2 footpad with a conservative adhesive pressure of 10^5 N/m^2 provides a 1000 nN anchorage force, allowing the nanorobot to remain adhered to the skin even in the strongest water currents and air currents. A finger rubbed hard against the skin with the objective of dislodging adhered epidermal-resident nanorobots may apply up to 10 N of force over a ~1

cm^2 area, giving a ~10^5 N/m^2 shear force (~1,000,000 nN) which may be resisted by ten 100 $micron^2$ footpads each having a maximum adhesive pressure of ~10^6 N/m^2.

What about epidermal penetration? Complex and subtle low-speed methods are readily imagined, but let us consider here a simple brute-force approach. Failure strength of human skin is ~10^7 N/m^2, so a 1 $micron^2$ rod may be driven through the skin with a force of ~10,000 nN. A 100-micron long telescoping manipulator arm with stiffness ~25,000 nN/nm can apply 10,000 nN to the rod at 1 cm/sec producing an elastic deflection in the arm of less than 1 nm, drawing ~100,000 pW of power at the moment of penetration.

The surfaces of toenails and fingernails have 2-20 micron features and longitudinal striations with 10-15 micron valleys, and are usually littered with dirt, bits of skin, and other surface debris as small as 1-2 microns in size. The nail surface is heavily keratinized, with a failure strength about 20 times higher than that of skin. Similar considerations apply as with the skin, in regard to nanorobot locomotion.

What happens if the nanorobot falls off, or is otherwise removed from, the epidermal surface? Solem has analysed the case of nanorobots walking on walls and ceilings. Setting the electrostatic image force equal to the gravity force $\rho_{nano} L_{nano}^3 g$ for a cubical conducting-surface nanorobot of edge L_{nano}, the voltage required to cling to the ceiling without falling off is:

$$V_{plate} = z_{sep} \left(\frac{2 \rho_{nano} L_{nano} g}{\varepsilon_0 \kappa_e} \right)^{1/2} (volts)$$

Taking z_{sep}=0.5 microns, ρ_{nano}=2000 kg/m^3, g=9.81 m/sec^2, ε_0=8.85×10^{-12} farad/m, κ_e=1, and L_{nano}=10 microns, then V_{plate} ~ 0.1 volt. Similarly, the largest nanorobot that can cling to the ceiling using an adhesive contact surface of strength S has a characteristic dimension:

$$L_{max} \sim \frac{S}{\rho_{nano} g}$$

For S=10-100 N/m^2, then L_{max} ~ 500–5000 microns. By wetting the contact area with water having surface tension γ ~ 73×10^{-3} N/m at room temperature, a cubical nanorobot clinging to the ceiling may have a maximum characteristic dimension of:

$$L_{max} \sim \left(\frac{4\gamma}{\rho_{nano}\, g}\right)^{1/2} \sim 3900\, microns$$

NANOFLIGHT

Perhaps surprisingly, aerobotics has many useful applications in nanomedicine. Examples include monitoring and diagnostic systems, since flying nanorobots can transit the length of a patient's body in ~1 sec or less; sterile fields; personal defensive systems; and accident recovery systems. Unfortunately, in 1998 the aerodynamics of micron-scale flight systems was only lightly studied and aerial robotics was still considered an exotic sport or a military curiosity.

Nanoflight and Reynolds Number

It has been observed that a 30-cm paper airplane will glide slowly and stably, but a 3-cm paper airplane made from thinner paper requires a much higher velocity/size ratio to remain airborne, and with a noticeable lack of stability – sometimes the plane flies well, sometimes not. If size is further reduced into the millimeter range, the plane almost cannot fly.

As in the case of swimming, this transition can be explained in terms of the Reynolds number N_R, the ratio of inertial to viscous forces acting on a body that is passing through a fluid such as air. Generally speaking, microscopic organisms or flying nanorobots with $N_R \ll 1$ move by utilizing viscosity, while macroscopic objects such as aircraft with $N_R \gg 1$ use inertia to generate lift. Millimeter-size airfoils with N_R ~0.1-100, as typified by flying insects, occupy a transitional regime. Insects make use of both inertial and viscous forces, often employing unusual wing flapping patterns and elastic energy storage systems to remain aloft. The smallest known flying insect that can make any use of aerodynamic lift forces is the four-winged parasitic chalcid wasp *Encarsia formosa,* which has a total wingspan of ~1.4 mm. Aerobotic machines with wingspans smaller than ~100 microns probably must make almost exclusive use of viscous propulsive forces.

A nanorobot of dimension L flying at velocity v through 20°C sea-level dry air (ρ_{air}=1.205 kg/m^3, absolute viscosity η_{air}=0.0183×10^{-3} kg/m-sec) has a Reynolds number N_R=66,000 v L. Viscous forces dominate when $N_R < 1$, or when:

$$v < \frac{15}{L_{micron}} \quad (m/sec)$$

where L_{micron} is characteristic nanorobot size expressed in microns. Thus a 1-micron nanorobot remains in the viscous regime up to ~ 15 m/sec flight velocity (34 mph), a sufficient speed for most medical applications. Note that formulas involving viscous forces may not apply to aerobotic airfoils with L <~ λ_{gas}, indicating transitional or ballistic flow.

The main implication of is that conventional aeronautics technologies such as rigid wings and jets are usually inappropriate for micron-size flyers. Instead, aerial nanorobots may more profitably employ natation mechanisms as described in, including surface deformations, inclined planes, volume displacement, and viscous anchoring. Specific aeromotive mechanisms are of great interest but will not be considered further in this Volume. In many cases, a viscous-regime nanorobot may exit the bodily fluids and enter the atmosphere,* or vice versa, using the same propulsive mechanisms, though operated at a modified speed or pitch angle.

Nanoflight and Gravity

In addition to forward progression through the medium, atmospheric flight requires active and continuous support against the pull of gravity. For small spherical objects of radius R=R_{nano} in the laminar flow (low N_R) regime, the rate of fall in a still medium is approximated by Stokes Law for Sedimentation. For nanorobots near the Earth's surface that are falling in air, with $g=9.81$ m/sec^2, $\rho_{particle}$~1000 kg/m^3, $\rho_{fluid}=\rho_{air}$ and $\eta=\eta_{air}$, then terminal velocity $v_t=1.2\times10^8\ R_{nano}{}^2$ (m/sec). An $R_{nano}=1$ micron nanorobot falls at $v_t=120$ microns/sec; an $R_{nano}=10$ micron nanorobot falls at $v_t=1.2$ cm/sec, requiring a rather modest power expenditure of only P_{nano}/e Percent=(0.5 pW)/e Percent to remain aloft.

Buoyant Nanoballoons

Gravity may also be overcome by reducing density relative to the surrounding medium. An object has neutral buoyancy when its density equals that of the medium in which it is suspended, becoming "weightless" within that medium. A nanorobot whose interior volume consists 90 Percent of vacuum has a v_t ~10 times smaller than a completely solid object of equal size.

What is the tiniest possible lighter-than-air balloon? J.S. Hall notes that for a one-atom-thick graphene shell the out-of-plane bending stiffness of the C-C bond is much lower than the in-plane stretching stiffness. This is why hollow fullerenes of submicron diameter are experimentally observed to collapse even when their interiors are not evacuated. Simple internal bracing sufficient to stabilize an evacuated structure outweighs the lift. Applying the Euler buckling formula to three diamondoid orthogonal diametral stiffening rods inside a spherical evacuated nanoballoon gives a scale-invariant total beam mass ~6 times the mass of the displaced air, hence net lift is impossible for any device radius using this minimal crossbeam design although macroscale geodesic trusswork-stabilized vacuum balloons cannot be ruled out.

Hall also observes that pressurizing nanoballoons to atmospheric pressure removes most shell stress. This strategy also eliminates the need to thicken the single-atom shell walls, up to a nanoballoon radius of at least $\sigma_w\ \tau_{wall}/\Delta p$ ~ 100 microns, taking wall thickness $t_{wall}=0.17$ nm, a conservative diamondoid wall working stress $\sigma_w=10^{10}$ N/m^2, and allowing a maximum environmental pressure fluctuation of Δp ~ 0.17 atm (vs. ~0.002 atm for 140 dB sound waves, ~0.1 atm normal barometric variation, and ~2 atm maximum sound pressure in air).

The smallest atmospheric-pressurized atomic-walled nanoballoon that can achieve neutral buoyancy has radius $R_{min}=3\ \rho_{wall}\ t_{wall} / (\rho_{air}-\rho_{gas})$, where ρ_{wall} is wall density, ρ_{air} is air density (ρ_{air}=1.2929 kg/m^3 for dry air at STP), and ρ_{gas} is the density of the filling gas. Thus R_{min}=1.6 micron for hydrogen gas with ρ_{gas}=0.0899 kg/m^3; R_{min}=1.7 micron for STP helium gas (r_{gas}=0.1785 kg/m^3) which, in conjunction with a slightly thicker nondiamondoid shell, eliminates flammability concerns at all device number densities. Diffusion leakage must also be addressed.

Nanoballoons of radius $R > R_{min}$ can carry payloads of mass:

$$M_{payload} = \frac{4}{3}\pi\,\rho_{air}\,R^3\left(\frac{\rho_{gas}}{\rho_{air}}\right) - 4\pi\,\rho_{wall}\,t_{wall}\,R^2$$

A helium-filled R=2.2-micron fullerene sphere can lift a ~10^{-17} kg payload mass, representing a payload volume of ~0.01 micron^3 at $r_{payload}$ ~ 1000 kg/m^3. Expanding nanoballoon radius to R=6.8 microns increases payload volume to ~1 micron^3. Pressurized buoyancy-based lift systems may be useful either in early-generation aerial nanodevices that must rely upon primitive energy supplies, or in default-float applications. However, the modest power expenditure needed to overcome gravity in micron-size devices suggests that nonbuoyant active-propulsion designs will normally be preferred.

Nanoflyer Force and Power Requirements

What forces must be applied, and what power must be expended, in order for a flying nanorobot to maintain continuous and controlled progression through the air? An exact calculation requires a detailed knowledge of the mode of locomotion, the shape and dimensions of wing or body, airfoil surface characteristics, and many other factors. However, the force needed to drive a spherical flyer of mass

m_{nano} through the air at a velocity v_{nano} may be approximated by summing the three drag components of skin friction ($F_{viscous}$), pressure drag ($F_{inertial}$), and induced drag due to lift ($F_{induced}$), as:

$$F_{nano} = F_{viscous} + F_{inertial} + F_{induced} \ (newtons)$$

$$F_{viscous} \sim 6\pi \eta_{air} R_{nano} v_{nano}$$

$$F_{inertial} \sim C_D \left(\pi R_{nano}^2\right)\left(\rho_{air} v_{nano}^2\right)/2$$

While conservatively taking the dimensionless coefficient of drag $C_D \sim 2$. Induced drag, associated with the shedding of transient vortexes from each wing tip, is an important consideration in macroscale winged vehicles but may be ignored here for microscale flyers operating in the viscous or transitional flight regimes.

The nanorobot aeromotive energy plant, of efficiency e Percent, must develop a continuous power of:

$$P_{nano} \sim \frac{v_{nano} F_{nano}}{e\%} \ (watts)$$

giving a whole-device power density of:

$$D_{nano} \sim \frac{3 P_{nano}}{4\pi R_{nano}^3} \ \left(watts/m^3\right)$$

These formulas give somewhat conservative results because they do not take into account special body shapes, dynamic wing motions, virtually complete pressure recovery in creeping flow, and so forth that may be used to improve flight efficiency in an optimized design. On the other hand, flight efficiency may be reduced by the accretion of water molecules and other environmental substances on working surfaces, and by the power expended in shearing air or in imparting net kinetic energy to the air.

Force, power, and power density for spherical flying nanorobots of various sizes and airspeeds, computed using the above relations. Note that power scales as $\sim v_{nano}^2$ in the viscous regime, $\sim v_{nano}^3$ in the transitional regime. The maximum sustainable velocity for transitional-regime insects is ~11 m/sec.

For an R_{nano}=1 micron aerial nanorobot, taking η_{air}=0.0183×10^{-3} kg/m-sec and ρ_{air}=1.205 kg/m^3 in dry air at 20°C and at 1 atm, with ρ_{nano}~1000 kg/m and e Percent=0.10 (10 Percent), a v_{nano}=1 cm/sec airspeed requires F_{nano}~4 pN and a P_{nano}~0.4 pW powerplant (D_{nano}~82,000 watts/m^3). At v_{nano}=1 m/sec, then F_{nano}~350 pN and P_{nano}~3500 pW powerplant (D_{nano}~8×10^8 watts/m^3). A much larger R_{nano}=1 mm nanorobot flying at v_{nano}=1 m/sec requires F_{nano}~4 microN and P_{nano}~41 microW (D_{nano}~10,000 watts/m^3). Again, an optimized design might reduce some of the power figures by a factor of 10 or more.

Aerobot power density D_{nano} scales as ~v_{nano}^{2} and ~R_{nano}^{-2} in the viscous regime. In the transitional regime, power density scales as ~v_{nano}^{3}, and as ~R_{nano}^{-2} at low velocities and ~R_{nano}^{-1} at high velocities. Thus, to minimize power density and therefore conserve energy, circumcorporeal clouds of aerial nanorobots may coalesce into progressively larger but fewer tightly-packed clumps as the collective velocity of the cloud moves to higher airspeeds, assuming that the aeromotive mechanism design is largely scale-invariant over the full size range of the progressive aggregations. This strategy is most effective in the viscous regime.

Consider a cloudlet consisting of 1 million nanorobots, each of size R_{nano}=1 micron. With individual nanorobots traveling at v_{nano} ~ 30 cm/sec, the cloudlet consumes ~0.4 milliwatts and operates at a power density of ~10^8 watts/m^3. Now assume that the cloudlet must speed up to 10 m/sec to track a fast-moving object around which it is stationkeeping, or to compensate for a heavy wind. If the individual nanorobots comprising the cloudlet simply increase their airspeed to 10 m/sec, then power density in each nanorobot increases to ~10^{11} watts/m^3 and cloudlet power consumption rises to ~400 milliwatts. However, if the nanorobots temporarily aggregate into a single collective approximating a single device of R_{nano}=100 microns, then the power consumption of the collective can be held to the original 0.4 milliwatts and power density remains constant at ~10^8 watts/m^3. Facultative aggregation may permit stationkeeping over a wide range of velocities without significantly increasing power. Other power-conserving behaviours, such as preferential migration into the downwind slipstream of a rapidly-moving tracked object, are not considered further here but may be useful.

How fast can nanoflyers accelerate? The answer is highly design-dependent but a crude generalization may be made. Consider a spherical nanorobot of radius R_{nano} that uses circumferential wings of similar size to impart momentum to a nearby mass of air by speeding up that mass of air to a higher velocity. The wings sweep air from a circular cross-section of radius $2R_{nano}$. The nanoflyer accelerates with constant acceleration $a_{nano}=v_{nano}^2 / 2 X_{accel}$ from a standing start to a final velocity v_{nano} in a time $t_{accel}=v_{nano}/a_{nano}$ and a running distance X_{accel}. The mass of the swept-out air is $M_{air}=4 \pi \rho_{air} X_{accel} R_{nano}^2$ and the mass of the nanorobot is $M_{nano}=(4/3) \pi \rho_{nano} R_{nano}^3$. To conserve momentum, $M_{air} v_{air}=M_{nano} v_{nano}$, hence $v_{air}=\rho_{nano} v_{nano} R_{nano} / 3 \rho_{air} X_{accel}$; $v_{air} < v_{sound}=343$ m/sec in dry air at 20°C and 1 atm to remain subsonic. The kinetic energy imparted to the sweptout air is $KE_{air}=(1/2) M_{air} v_{air}^2$ and to the nanorobot is $KE_{nano}=(1/2) M_{nano} v_{nano}^2$, with $KE_{total}=KE_{air} + KE_{nano}$. Neglecting drag losses and assuming negligible losses within the air-accelerating mechanism itself and any drag on the air used as reaction mass, then propulsive efficiency is e Percent ~ KE_{nano} / KE_{total} and total power consumption is $P_{nano} \sim P_{drag} + (KE_{total} /$ e Percent $t_{accel})$. In the examples below we take $R_{nano}=1$ micron, $v_{nano}=1$ m/sec, $P_{drag} \sim 3500$ pW, $\rho_{air}=1.205$ kg/m^3 for dry air at 20°C and 1 atm, and $\rho_{nano}=1000$ kg/m^3; here the Reynolds number is near unity.

A 1 micron^3 high-density powerplant develops $P_{nano} \sim 1$ million pW. At this high power level, a nanorobot can accelerate at $a_{nano} \sim 12{,}000$ g's for $t_{accel} \sim 9$ microsec, reaching $v_{nano}=1$ m/sec after crossing a running distance of $X_{accel} \sim 4.4$ microns with an efficiency of e Percent ~ (0.016) 1.6 Percent and $v_{air} \sim 64$ m/sec. As an alternative approach, note that a 5-microsec gas discharge from a simple pressure-release pump of volume $v_{reservoir} \sim 0.1$ micron^3 having a pressure differential of 1000 atm produces a mean discharge power of P_{nano} ~ 2 million pW, which allows a_{nano} ~ 15,000 g's of acceleration with e Percent ~ (0.01) 1 Percent for a maximum gas exit velocity of v_{max} ~ 70 m/sec using a nozzle of length l_{tube}=1 micron and radius r_{tube} ~ 10 nm.

At a more modest P_{nano} ~ 4500 pW, the nanoflyer accelerates at a_{nano} ~ 1100 g's for t_{accel} ~ 96 microsec, reaching v_{nano}=1 m/sec after crossing a running distance of X_{accel} ~ 48 microns with an efficiency of e Percent ~ 0.15 and v_{air} ~ 6 m/sec.

Attitude control of flying machines is a problem often mentioned in MEMS aerobotics work; airborne nanorobots may employ gravity-based dynamic stabilization, gyrostabilization, or other means. In 1998, the energetics and control of steering maneuvers in highly-maneuverable insects was being investigated.

8

Other Capabilities

NANOCHRONOMETRY

Nanorobots will use clocks in many applications. Computer gating,* navigation, and high-speed sensor applications may require repeatable timing in the 1-1000 nanosec range, or 1-1000 microsec for lower-speed chemical and chemotactic sensing, although long-term clock stability is not especially critical for computing. Neuron-mediated signals and muscle motions are monitored and controlled on a 1-1000 millisec timescale, while conscious human action and most human biorhythms occur in the 1-10^5 sec range.

This Section briefly introduces human chronobiology, then describes possible nanoscale oscillator systems that could be useful in nanorobotic clocks, basic principles of pre- and post-infusion nanorobot chronometer synchronization, and dedicated chronometer organs.

Human Chronobiology

The human body incorporates numerous biological clocks. The best-known and most-studied example is the daily endogenous circadian oscillator. This internal clock is normally reset by natural sunlight, which is much brighter than indoor lighting. The clock, in turn, sets the cadence for most of the other 24-hour body rhythms — sleep/wakefulness cycles, urine production, body temperature, blood cortisol and ACTH cycles, and the diurnal rhythm of mitosis in epidermal epithelium. The circadian clock stability is ~1 Percent in *Drosophila* and ~0.2 Percent in humans.

Mammalian circadian rhythms are regulated by a master pacemaker within the suprachiasmatic nuclei of the hypothalamus. In humans, the SCN is comprised of ~10,000 special cells that send out electrochemical signals in a 24-hour timing pattern. Entrainment of the clock to light-dark cycles is mediated by photoreceptors in the retina, with light information conveyed directly from ganglion cells of the retina to the SCN via the retinohypothalamic tract. Mammals and other vertebrates have another independent circadian clock in each retina, possibly driven by the retinal cryptochromes CRY1 and CRY2, producing rhythms in local physiologysuch as the diurnal renewal cycles of rods and cones. CRY1 is also abundant in skin tissues. Clock-resetting photoreceptors have been found in the human popliteal region, and such photoreceptors may exist in many different tissues throughout the human body.

The current model for a core circadian oscillator comprises, in part, a transcription/translation-based negative feedback loop in which clock genes are rhythmically expressed, giving rise to cycling levels of clock RNAs and proteins. The proteins then feed back, after a lag, to depress the level of their own transcripts, perhaps by interfering with positive elements that increase transcription of the clock genes. In 1998 this genetic network was still in the discovery process. One ~100,000-base-pair gene, aptly named "clock," was known to produce an 855-residue protein that serves as a major regulator of the ~10 other, still unidentified, genes thought to affect circadian rhythm. Circadian rhythms will almost certainly be made up of many interconnected feedback loops that interact with the CLOCK-related core pathway. It is believed that this interconnected ensemble will ultimately determine all of the classical circadian properties – period length, temperature compensation, and resetting by light or temperature. In 1998, most chronobiologists believed that many of these outer loops would be organism specific and that only the core loop would be more universal.

In addition to the 24-hour circadian oscillator, many other biological clocks have been identified in humans. The 28-day menstrual cycle is timed by two ovarian clocks, each delimiting a period of 14 days. One of these two clocks is the Graafian follicle and its production of estradiol, while the other clock is the corpus luteum and its secrction of progesterone. Both are obligatorily dependent upon the proper functioning of a third clock located in the arcuate region of the hypothalamus, known as the GnRH pulse generator, which ensures the rhythmic production and release of the gonadotropic hormones into the peripheral circulation with a cyclical period of ~1 hour (±25 Percent stability), coincident with a 1-hour cycle of hypothalamic electrical activity (±12 Percent stability). A 90-minute clock paces the development of the somites, blocks of tissue that form in regular arrays along the spinal cord of vertebrate embryos. One regulatory gene, named "chairy," undergoes repeated 90-minute activity cycles and is known not to require protein synthesis; gene expression follows the repeating pattern even when protein synthesis is biochemically blocked. A much longer half-weekly pattern has been observed in mitotic activity in cancer patients.

Other biological clocks abound. The stomach produces rhythmic contractions of ~3/minute during digestion. Rhythmic segmentation contractions of the bowel at ~0.2 Hz are driven by a thin layer of specialized pacemaker cells called the intestinal cells of Cajal, which demonstrate electrical oscillations at about the same rate as the contractions. Cardiac pacemaker cells in the sinoatrial node emit a cyclical pulse that triggers a heartbeat at a ~1 Hz basal rate. The respiratory centres in the medulla oblongata produce an autonomic contraction cycle of the diaphragm at ~0.3 Hz. The human eye produces many spontaneous rhythmic dilations and constrictions, including high-frequency ocular microtremors present in all people, accommodation microfluctuations (~1-2 Hz) with the pupil varying <300 microns in diameter, and low-frequency pupillary oscillations. Electrical rhythms generated by the brain that are detectable by an electroencephalograph include alpha waves predominantly in the occipital region at

8-13 Hz, beta waves in the frontal and central areas at 18-30 Hz, delta waves at <4 Hz during deep sleep or abnormal function, and theta waves in the temporal and parietal areas at 4-7 Hz.

Individual cells display a number of oscillatory biochemical pathways as well, including calcium ion oscillations, cyclic AMP signalling, and various other cell cycles. The glycolytic enzyme oscillator, controlled primarily by phosphofructokinase, may mediate various rhythmic physiological behaviours such as slow waves of contraction in smooth muscle, electrical activity in neurons, and insulin release from β islet cells of the pancreas. In one experiment, this oscillator mediated a 55 mV voltage change in guinea pig cardiomyocytes at 0.010 Hz (±10 Percent stability) in an almost perfectly sinusoidal pattern, although a few cells displayed oscillations with irregular phase, amplitude, or both. In another experiment, fluorescently-labeled puffs of tumor-cell cytoplasm were observed being released through plasma membrane ruptures, regularly at ~0.05 Hz during a neutrophil-mediated cytolytic attack, matching the nicotinamide-adenine dinucleotide phosphate and superoxide release periods for neutrophils. By 1998 there was a growing belief that many, if not most, of the cells in an animal may possess individual biological clocks; fruit flies were already known to have clocks distributed throughout their wings, legs, and abdomen.

The human brain also possesses an interval timer that allows a person to gauge the passage of seconds or minutes to a mean accuracy of ±15 Percent. The interval timer functions like a stopwatch and resides in the basal ganglia, a region of the brain that coordinates voluntary muscle movements. A population of neurons in the substantia nigra releases regular pulses of dopamine into an accumulator, called the caudate-putamen, a major part of the basal ganglia. These pulses are then read via neural pathways to the cerebral cortex which are known as striato-cortical loops. Other internal "alarm clocks" that reliably allow waking from sleep at desired times* have been described.

In vivo nanorobots should be able to eavesdrop on most of these active chronobiological channels, gaining complete knowledge of the oscillator frequencies and phase settings of virtually all of the human body's biological clock systems. Blood plasma cortisol typically peaks in the morning, then declines throughout the day. Melatonin cycles in the opposite direction, normally peaking at $6\text{-}7\times10^{-11}$ gm/cm^3 at night and then falling to 1.4×10^{-11} gm/cm^3 during the day. Longer cycles are easily tracked too. Blood viscosity in healthy young women shows the rhythmic variation of the menstrual cycle, probably due to changes in serum fibrinogen and globulin levels which can be measured more directly. These nanorobot-accessible readings may be used either to measure "body time" or to control it, and in some cases can recalibrate relevant onboard nanochronometers.

Environmental time cues may also be detected by in vivo nanorobots, such as the level of illumination penetrating the epidermis or the all-pervasive 60/50 Hz electromagnetic hum generated by alternating current electrical supply systems in common use throughout the world, a crude frequency standard whose voltage can sometimes display time-of-day load-related fluctuations.

NANOROBOT SYNCHRONIZATION

Ideally, the clocks of all nanorobots present in a patient's body will be synchronized to some universal time. This time setting may be defined by the patient, by the physician, or by reference to some external standard. As part of a chronometry-critical nanorobot installation procedure prior to infusion, an initializing signal can be transmitted throughout the infusate volume, allowing each individual nanorobot present therein to receive the signal and set its clock.

Acoustic synchronization pulses are appropriate if ~microsec precision will suffice. Sound waves passing at v_{sound}=1500 m/sec through an L_{vial}=1.5 cm wide container of well-stirred aqueous-suspended nanorobots restricts synchronization error to $\Delta t_{error} \sim L_{vial}/v_{sound}$=10 microsec, repetition of N_{obs}=100 synchronization pulses can reduce synchronization error to ~1 microsec.

For the highest pre-infusion synchronization accuracy, optical or rf pulses will be employed. Electromagnetic energy travels at c ~ 3 x 10^8 m/sec across the container volume until intercepted by a nanorobot optical sensor, giving $\Delta t_{error} \sim L_{vial}$ / c=0.05 nsec. Allowing 100 green photons (360 zJ/photon at 550 nm) per nanorobot at a nanorobot number density of 10^{12} cm^{-3}, the flash injects ~40 J/m^3 of energy into the container, raising the water temperature by ~10 microkelvins after this energy is fully thermalized. A 1-nsec flash produces a peak intensity of ~4×10^{10} watts/m^2, comparable to monochromatic optical tweezers which are tolerated by biological macromolecules. Some nanorobots inevitably will miss the signal and may later synchronize, prior to infusion, via close or direct physical contact with others who received the signal, or else multiple synchronization pulses can be used.

Onboard nanorobot clocks can also be synchronized post-infusion. A pressure cuff inflated around the patient's arm can administer acoustic synchronization pulses to ~microsec precision as bloodborne nanorobots pass through the blood vessels within; after signaling for several mean blood circulation times, most bloodborne nanorobots should be properly synchronized. Whole-body acoustic transmissions at 0.1-1 MHz may be generated by operating tables, vibrating chairpads, wristwatch transmitters and the like. Such methods may produce synchronization errors of ~100 microsec over 15 cm anteroposterior path lengths, reducible to ~1 microsec error using N_{obs}=10,000 repetitions of the calibration signal. For higher precision, the physician transmits ~MHz radio wave pulses into the body; such pulses may be detected by appropriate onboard receivers. Such waves have only ~75 Percent attenuation over 15-cm anteroposterior path lengths, producing 0.5 nsec synchronization accuracy. Optical photons applied to the skin surface are subject to scattering, producing signal pulse broadening which increases synchronization error, and to absorption, which makes signal reception difficult or impossible beyond tissue depths of a few centimeters.

Another approach is to inject a relatively small number of chronocytes – mobile communicytes modified to include an onboard nanoclock of high precision and the ability to transmit clocking synchronization signals to calibrate neighbouring nanorobots. These signals may be transmitted acoustically at close proximity to the nodes of a mobile communications network, subsequently allowing all nanorobots within 100 microns of a calibrated node to synchronize to within <~67 nsec. Averaging algorithms can be employed to synchronize uncalibrated nodes in the system that were not recently visited by a chronocyte. Bloodborne chronocytes will normally be carried from the capillary vasculature to within ~100 microns of most tissue locations, thus allowing <~67 nsec single-pulse acoustic synchronization even in the absence of an installed communications network.

DEDICATED CHRONOMETER ORGANS

Just as the hypothalamus is the physical locus of the natural circadian clock, it may be convenient to establish artificial chronometer organs inside the human body for various purposes. Dedicated nano-organs have been described previously in connection with power, communication, and navigational systems.

Chronometer organs, which would likely be millimeter-scale or smaller, could be used as endogenous clocks to regulate the precise administration of time-release substances or devices. Such organs could serve as highly accurate timers to improve the natural human interval timing sense from ±15 Percent accuracy in biological systems to parts per million accuracy, or better, using artificial systems. If linked to the patient or user through various outmessaging channels, chronometer organs can provide a continuously-available, consciously-accessible unfailing "time sense" of extraordinary precision for time of day, calendar date, and other time-related information.

Dedicated chronometer organs can be used to synchronize chronocytes or mobile communication networks at the nodes.

Chronocytes with lower-stability crystal oscillators could be repeatedly recalibrated as they passed near larger embedded chronometer organs during each blood circulation cycle. Such chronometer organs might incorporate higher-stability onboard atomic frequency standards. Alternatively, chronometer organs could serve as transdermal data ports through which timing synchronization signals can be injected into fibre-based in vivo communication networks from external timing sources, similar to the autoclock pulse that is added to most contemporary television broadcast signals. Hard connectors, centimeter-scale rf wireless antennas, and other types of links are possible. In 1998, a desk clock could be purchased for $50 that automatically recalibrated itself by picking up WWVB radio signals anywhere in North America, several times a night, using a few centimeters of rod antenna. WWV's carrier frequency is regulated to Dn / n ~ 5 x 10^{-12}, ultimately providing a time synchronicity at the receiver of ~100 microsec/day. With a slightly larger in vivo antenna, satellite GPS signals may be received, providing <~20 nanosec time error. If buildings and vehicles are rewired to permit dissemination of time synchronization pulses and other useful information via continuous rf or IR channels, chronometer organs could employ smaller in vivo antennas and still achieve very precise results.

Although it has been informally speculated that nanorobots could use biological nerve fibres as rf antennas to receive electromagnetic time recalibration signals from outside the human body, such as from WWV, this concept is probably unworkable for several reasons. First, the axoplasm is only ~10^{-7} as electrically conductive as a metal wire of equivalent size, because axoplasmic charge carrier density and mobility are much lower than for electrons in a wire. Second, passively conducted axonal currents are quickly attenuated by leakage through ion channels in the membrane, which is a very poor insulator. From cable theory applied to neurons, the length constant λ_n is defined as the distance over which an applied potential depolarizes to 1/e (~37 Percent) of its maximum value; $\lambda_n=(d_{neuron}\ r_{mem}/4\ r_{axo})^{1/2} \sim (0.04\ d_{neuron})^{1/2}$, where

d_{neuron} is axonal diameter in meters, r_{mem} is the specific membrane resistance (~0.2 ohm-m^2 in human neurons) and r_{axo} is the specific resistance of the axoplasm (~1.25 ohm–m in human neurons). For a typical human axon with d_{neuron} ~ 1 micron, then l_n ~ 200 microns (~internodal distance between nodes of Ranvier) and an external signal is >99 Percent attenuated in ~1 mm of longitudinal travel. Third, short pulses are severely distorted and attenuated by the electrical capacitance of the cell membrane. The capacitive time constant in human nerves and muscle cells ranges from 120 millisec, thus limiting any possible electromagnetic reception to frequencies under 50-200 Hz, although pulsed microwaves directed through the brain can induce auditory effects in animals and humans.

NANOCOMPUTERS

Many important medical nanorobotic tasks will require computation during the acquisition and processing of sensor data, the control of tools, manipulators, and motility systems, navigation and communication, and during the coordination of collective activities with neighbouring nanorobots. Ex vivo computation has few theoretical limits, but computation by in vivo nanorobots will be subject to a number of constraints such as physical size, power consumption, onboard memory and processing speed.

The memory required onboard a medical nanorobot will be strongly mission dependent. Recognizing and manipulating molecules is fundamental. A very simple mission might demand only the identification or handling of perhaps ~10 different molecules. Basic respiratory gas transport nanorobots such as respirocytes may require the operation of fixed-shape receptors that bind simple molecules such as O_2, CO_2, H_2O, and glucose. Identifiers for such receptors may need only a few bits, hence this memory requirement should be negligible. A toxin removal device similarly may require keeping track of only a few types of fixed-shape receptors. On the other hand, a survey or assay mission might need to recognize N=100-1000 distinct proteins. A spherical 1-micron nanorobot can have $>10^4$ fixed-shape receptors on its surface; if these will suffice, then we require N $\log_2$ (N) ~ 10^4 bits to identify each of N=1000 different receptor types. This seems more efficient than using more advanced reconfigurable receptors, which might need $>10^4$ bits per receptor-pattern to specify each binding site geometry to the necessary atomic-scale resolution, thus imposing a total memory requirement of $>10^7$ bits for an onboard library of N=1000 different receptor types. Consequently, simple missions involving basic process control with limited motility may require no more than ~10^5-10^6 bits of memory, comparable to an old Apple II computer (including RAM plus floppy disk drive). At the other extreme, a complex cell repair mission might require the onboard storage of a substantial fraction of the patient's genetic code, representing ~10^9 bits of memory including perhaps ~0.2×10^9 bits of linear sequence data for all 100,000 protein types found in the human body, again assuming 300 amino acids per protein. Most amino acids are folded into the protein's interior and are not readily accessible to surface probing unless the protein is unfolded, which usually is not desirable or convenient. On the other hand, binding sites for large molecules should be physically easier to construct than small-molecule receptors. An onboard memory of 10^9-10^{10} bits would be in the same range as the 1985 Cray-2 (2×10^{10} bits) or the 1989 Cray-3 (6×10^8 bits) supercomputers.

Computational speed will also be strongly mission dependent. However, extremely simple process control systems in basic factory settings may only require speeds as slow as 10^4 bit/sec. Individual natural biocomputational devices generally do not exceed this speed. Examples include mRNA translation during protein manufacture at ~15 Hz (~75 bits/sec assuming 5 bits/protein); transcription from DNA by RNA polymerase at ~40 Hz (~80 bits/sec at 2 bits/nucleotide); DNA replication at ~800 Hz (~1600 bits/sec at 2 bits/nucleotide); typically 5-100 Hz (bits/sec) for neural electrical discharges ~1000 Hz (bits/sec) for excitory cholinergic synapses, and gated ion channels at ~10^4 Hz (bits/sec). At the other extreme, a processing speed of 10^9 bits/sec allows a ~10^9

bit genomic information store to be processed in ~1 sec, the small-molecule diffusion time across an average 20-micron wide cell. In 1998, personal desktop computers capable of ~10^9-10^{10} bits/sec were commonly available.

Nanomechanical Computers

Electronic computers were evolutionarily preceded by purely mechanical computational devices, starting with the venerable abacus in ~3000 BC, the Pascaline in 1642, and the Difference Engine designed in 1821 by Charles Babbage. A 2000-part working subsection of the brass-geared Engine was demonstrated in 1832; an entire working Difference Engine was reconstructed by historians in 1991, proving that Babbage's design was sound. In the 1840s, Thomas Fowler built and exhibited a calculating device using sliding rods made of wood instead of metal. Whereas Babbage's engines used the familiar 0-9 decimal system with each number represented by a discrete position of a rotating gear wheel, Fowler's machine was more fully digital, using as its active element not rotating wheels but sliding "trinary" rods which could occupy only one of three positions at any time, the first known example of "rod logic."

By 1834, Babbage had also conceived detailed plans for his Analytical Engine, intended as a general-purpose programmable computing machine but based entirely on 19th century mechanical technology. The Analytical Engine was to have a random-access memory consisting of 1000 words of 50 decimal digits each (~175,000 bits), with separate memory and central processing unit (CPU), stored programme control, data entry via punched metal cards, and even an output printer. This ambitious device, though well specified, was never built.

The mechanical computing tradition was not entirely abandoned. Vannevar Bush built his analog mechanical computer, the Differential Analyser, in 1930 at MIT. In 1954, M. Minsky and R. Silver used hydraulic logic elements to build a mechanical "hydroflip computer" which was operated at ~30 Hz and powered by a 3-inch high column of water. In 1975, D. Hillis and B. Silverman built a special-purpose all-mechanical computer ~2 meters in size, entirely out of Tinkertoys, and powered by a hand crank, that was able to play tic-tac-toe. In 1990, University of Minnesota engineers fabricated a complete family of micromechanical digital logic devices, including electrostatically-actuated linear-sliding ~30 micron mechanical logic elements confined to a one-dimensional track, forming NAND and NOR gates suitable for low-speed radiation-hard digital functions "in environments hostile to electronic devices." Sandia's pin-in-maze microlocks could also be used to make a mechanical computer, albeit at the above-micron scale. In 1996, J. Gimzewski and colleagues at IBM Zurich used a scanning tunneling microscope (STM) probe to repeatedly reposition spherical C_{60} fullerene molecules ~1 nm in diameter along a terraced copper substrate that constrained the buckyballs to move only in a straight line, operating the mechanical array like beads on an abacus. In 1997, Stoddart's group described a mechanical XOR gate based on their "molecular shuttles."

Perhaps the best-characterized mechanical nanocomputer is Drexler's rod logic design. In this design, one sliding rod with a knob intersects a second knobbed sliding rod at right angles to the first. Depending upon the position of the first rod, the second may be free to move, or unable to move. This simple blocking interaction serves as the basis for logical operations. A nanomechanical implementation of a Boolean NAND "interlock" gate, using clock-driven input and output logic rods 1 nm wide which interact via knobs that prevent or enable motion, all encased in a housing, allowing ~16 nm^3/ interlock. The activity sequence depicts a simplified version of Drexler's register, omitting the reading mechanism. Initially the register shows a 0 or a 1 (B). Then the barrier is lowered and the ball wanders freely (C); entropy increases. The register is then reset to 0 by spring-rod compression, converting ~kT ln(2) joules of work into heat (D). To write a 0, the barrier is raised (E). To write a 1, the input rod at right is first extended and then the barrier is raised (F); the input rod does work compressing the ball into the spring, but this energy can be retrieved when the spring rod is retracted. Finally, the spring rod is retracted, returning the device to state (A) or (B).

A programmable logic array (PLA) using knobbed rods (omitting several drive and spring systems for clarity) to implement nanomechanical logic, which, in combination with rod-based registers, can be used to build much of the control circuitry for a CPU in a >~1 GHz clocked computer system. While a PLA system requires three successive rod displacement cycles to compute a set of Boolean functions, the functions AND, OR, and NOT can also be computed in a single displacement cycle by employing a linkage in which any input rod can displace the output rod without displacing any other input rod.

Drexler's benchmark mechanical nanocomputer design has 10^6 interlock gates, 10^5 logic rods, 10^4 registers, an energy-buffering flywheel and other components with total cubic volume (~400 nm)^3, mass ~10^{-16} kg, and total power ~ 60 nW, giving a power density of ~10^{12} watts/m^3. Power dissipation per logic operation is ~0.013 zJ per gate per cycle, giving (including register dissipation) ~2×10^4 operations/sec-pW. Processing speed is ~10^9 operations/sec (~1 gigaflop) or ~10^{28} operations/sec-m^3; assuming one bit per register, processing speed is ~10^{13} bits/sec or ~10^{32} bits/sec-m^3. In 1998, the typical desktop PC operated at ~10^8 operations/sec. Cooling is provided by a refrigerant fluid flowing through an integral fractal plumbing system at near the speed of sound.

The most primitive 4-bit Intel 4004 microprocessor, introduced in 1971 for early-generation pocket calculators, had 2300 transistors. A comparably simple early mechanical nanocomputer with 2000 interlock gates, 100 registers, and a 1 KHz clock speed could have a volume as small as 36,000 nm^3, a (~33 nm)^3 cube, assuming 16 nm^3/gate and 40 nm^3/register, and could process ~10^5 bits/sec.

One major disadvantage of mechanical nanocomputers is that they will almost certainly be slower than nanoelectronic systems, because a device that depends on moving heavy nuclei (~10^{-27} kg) will necessarily be slower than a device that depends on moving less-massive electrons (~10^{-30} kg). Mechanical signals travel near the speed of sound (~10^4 m/sec in diamond), whereas electronic signals may travel near the speed of light (~10^8 m/sec). Thus while nanomechanical computers may be limited to switching speeds of ~50 picoseconds, electronic devices may be ~10^3-10^4 times faster. On the other hand, mechanical computers are more EMP (electromagnetic pulse) resistant, conceptually easier to understand and to model, and mechanical designs scale readily from the macroscale to the nanoscale, unlike most nanoelectronic designs. In 1996, a ~50 nm reversible mechanical latch was fabricated and physically cycled using an AFM; the similarity to Drexler's mechanical OR gate was explicitly noted in the paper.

What about mechanical data storage? In the MEMS world, Halg fabricated a nonvolatile microelectromechanical memory cell consisting of a longitudinally-stressed surface-micromachined ~10 micron bridge that mechanically buckles into one of two states, making a bistable storage device. Moving to the nanoscale, data could be mechanically stored using compact three-dimensional arrays of diamondoid register rods. Another theoretical mechanical storage medium is spooled hydrofluorocarbon memory tape, or polymer or diamond surface-bound patterns of H and F atoms that could be read using an optimized $(CH_3)_3PO$ scanning probe tip with a maximum raw error rate of 6×10^{-8} per read.Direct AFM reading and recording of 10-nm pits on a polycarbonate surface and other means of directly reading and writing data at the atomic level have been studied, and the fullerene abacus described earlier is one of the first experimental efforts to mechanically store numerical information using individual molecules at room temperature.

It has been speculated that individual atoms or atomic vacancies could serve as information units. Y. Mo used an STM to store information in the reversible rotational states of individual antimony dimers deposited on a silicon substrate. In theory, a block of diamond with data encoded as single-atom lattice vacancies could store up to ~176 bits/nm^3, or nearly ~2×10^{11} bits/$micron^3$, although cryogenic temperatures might be required to reduce diffusion effects and structural lability to acceptable levels consistent with long-term information storage. Phonon probes or other nondestructive means of lattice interrogation could be employed for readout.

Nanoelectronic Computers

In 1998, the metal-oxide-semiconductor field-effect transistor (MOSFET) was still the most common type of transistor in general use. These solid-state devices are characterized by an electrical source, an electrical drain, and a gate that controls the flow. Since its inception, the FET has scaled well to lower sizes. Individual working transistors with 40-nm gate lengths have been demonstrated in silicon, and 25-nm gate-length transistors have been fabricated in gallium arsenide. However, as feature sizes shrink below <~100 nm, a number of obstacles to stable operation begin to appear and the cost-effective downscaling of dense circuitry may not persist. A number of solid-state replacements for the bulk-effect semiconductor transistor have been suggested that could overcome these obstacles by taking advantage of nanometer-scale quantum mechanical effects. Such replacements include quantum dots (QDs), resonant tunneling devices (RTDs), electron turnstiles, and single-electron transistors (SETs). However, while hybrid RTD-FET circuits and SETs have been successfully switched at room temperature, many of these devices must be operated at cryogenic temperatures and other practical challenges remain; such top-down approaches to nanoelectronics are not considered further here.

In contrast, molecular nanoelectronics uses primarily covalently bonded molecular structures, electrically isolated from a bulk substrate, producing wires, switches and devices composed of individual molecules and nanometer-scale supramolecular structures. While it is relatively difficult and expensive to sculpt trillions of identical nanoscale structures in bulk materials, individual molecules are natural nanometer-scale structures that can be manufactured to be exactly the same in near-mole quantities. The great versatility of organic chemistry offers more options for designing and fabricating nanoelectronic devices than are available in silicon. Investigators are designing, modeling, fabricating, and testing individual molecules and nanometer-scale supramolecular structures that act as electrical switches and even exhibit some of the same properties as small solid-state transistors. Three and four terminal devices with ~0.3-nm feature sizes have been examined computationally.

Molecular electronics is a rapidly growing area in the literature. J.M. Tour expects experimental demonstration of molecular-sized transistor devices by 2000-2001, and commercial high-level computers using molecular electronics by 2008-2013. A current list of major research groups is maintained by the International Society for Molecular Electronics and Biocomputing. Some of the following discussion is appreciatively drawn from an extensive 1997 review by MITRE Corporation.

BIOCOMPUTERS

Biocomputers use natural organic materials, thus offer the possibility of self-assembly or construction by natural or engineered biological organisms. At least three classes of biocomputers are distinguishable – biochemical, biomechanical, and bioelectronic.

Biochemical Computers

The Systron-Donner Analog Computer familiar to engineering students in the 1960s and 1970s used potentiometers and high-gain DC amplifiers, called operational amplifiers, to allow the creation of electrical circuits comprised of adders, multipliers, and integrators arranged in feedback loops capable of solving complex time-dependent differential equations modeling actual physical systems. In 1998, Motorola and others were producing programmable analog chips containing arrays of operational amplifiers and other analog components. Related control systems are commonplace in human physiology.

Biochemical feedback loops exemplified by enzyme activity in the living cell exhibit similar analog computational functionality. To start with, the net flow of carbon through an enzyme-catalyzed reaction may be influenced by

1. The absolute quantity of enzyme present,
2. The pool of nonenzyme reactants and products available, and
3. The catalytic efficiency of the enzyme.

All three influences may themselves be subject to recursive enzymatic control. The absolute quantity of enzyme present is determined by competing rates of synthesis and degradation; both of these processes are controlled by other enzymes. Product synthesis may be triggered by an inducer, or may be repressed by an end product. Local concentrations of coenzymes or metal ions can regulate catalytic efficiency. Inhibited activity of an enzyme in a biosynthetic pathway by an end product of that pathway, such as the much-studied inhibition of aspartate transcarbamoylase by cytidine triphosphate (CTP), is called feedback inhibition. Typically the feedback inhibitor, acting as a negative allosteric effector, binds to the sensitive enzyme at an allosteric site that is spatially distinct from the catalytic site. Further fine control is provided by additional multiple feedback loops, exhibiting such features as cumulative feedback inhibition, multivalent feedback inhibition, and cooperative feedback inhibition. Similarly, parallel chemical computers consisting of open, bistable coupled reaction systems capable of pattern storage and pattern recognition have been proposed and simulated to model a neural network, and experiments have been performed with a system of 16 coupled bistable chemical oscillators using the chlorite-iodide reaction.

Networks of interacting enzymatic reactivities can also be used to create biochemical digital computers. An enzymatic Turing machine was first proposed by C.H. Bennett, and Liberman described the living cell as an analog-digital molecular stochastic parallel computer. More recently, T. Knight of the MIT Artificial Intelligence Laboratory suggested that normal cellular activities can be coopted to implement digital logic gates, a process he calls "cellular gate technology." In this scheme, protein concentrations are used as binary signals. The signal is "1" if the marker protein is being synthesized, thus reaching some detectable threshold equilibrium value where the rates of synthesis and degradation are equal. The signal is "0" if the synthesis of the marker protein is being inhibited by some other protein, and thus the concentration is trending toward zero.

This principle may be used to make logical circuit elements. An inverter system consists of two different proteins, one of which is inhibiting the synthesis of the other. A NOR gate requires a system in which either of two proteins can inhibit the synthesis of a third protein; multivalent feedback inhibition in which two proteins are both required to inhibit production of a third makes a NAND gate. This is sufficient to implement any Boolean function, given enough gates and signals. Two NORs crossed in a loop makes a bistable multivibrator, or flip-flop, a fundamental component in data registers, shift registers, rectangular pulse generators, or pulse delay elements. The logic circuitry (the arrangement of "gates" and "wires") comprises a set of designed DNA sequences that determine which proteins are being synthesized and which combinations of proteins can inhibit the synthesis of others.

In the most optimistic case, we assume operation near concentrations of $c_{prot} \sim 10^{-4}$ gm/cm^3, marker proteins and gating enzymes of mass $MW_{prot} \sim 10$ kilodaltons, and a minimum of $N_{prot} \sim 100$ protein molecules per gate, giving a typical logic element size of $V_{gate} \sim N_{prot}\ MW_{prot}\ /\ N_A\ c_{prot}$ ~ 0.02 micron^3/gate where N_A is Avogadro's number. Flip-flop registers may store ~50 bits/micron^3. In theory, the DNA specifications for this computer could be grafted onto the genome of an animal cell or a bacterial cell, taking care to avoid any unwanted interactions with natural biochemical pathways. Minimum diffusion-limited gating time across a biochemical gate of characteristic dimension $DX \sim V_{gate}{}^{1/3}$ is given by Eqn. as $t = DX^2\ /\ 2\ D \sim 0.3$ millisec assuming protein marker diffusivity $D \sim 10^{-10}$ m^2/sec, consistent with ~KHz maximum operating frequencies when DNA is localized near the gates. For nucleoplasmic DNA with cytoplasmic computation, diffusion-limited maximum switching frequency falls to ~0.1-1 Hz. For comparison with natural systems, genes controlling the fibroblast cell proliferation programme have been observed to switch states in ~900 sec, implying computation at ~0.001 Hz.

Other classes of biochemically-modulated biomolecular switches have been investigated. Matthews incorporated an artificial molecular onoff switch into an enzyme by adding a pair of thiol groups into the active site of native lysozyme, replacing two amino acids on opposite side of the active site with cysteine, an amino acid with a thiol-containing side chain. Thiols may form covalent bridges (-SS-) or broken bridges (-SH HS-) under suitable chemical conditions. Cycling the solution composition causes the cysteines to form a bridge across the active site or to break the bridge; the process is completely reversible. Another experiment involved genetic modification to the bacterial phage lambda, a virus with two behavioural states called lytic and lysogenic. The NIH researchers installed a synthetic molecular switch in lambda that allows the virus to be biochemically toggled between its lysogenic and lytic phases by the introduction of active HIV-1 protease. The genetic circuit that controls the phage lambda lysis-lysogeny decision has been flowcharted in detail using Boolean logic operators. The immunological synapse – the specialized junction between a T cell and an antigen-presenting cell – acts as a kind of biochemical logic gate with a ~0.06 Hz switching rate.

In 1998, the DNA molecule also was being actively investigated for biochemical computers. DNA-based Turing machines were first proposed by C.H. Bennett and R. Landauer. In 1994, L. Adelman demonstrated the first DNA computer using fragments of DNA to compute the solution to a simple graph theory problem. Adleman used short DNA sequences to represent vertices of a network, or graph. Combinations of these sequences were then synthesized randomly by massively parallel chemical reactions in aqueous solution using a combination of the copying and combination reactions applied to the artificial DNA strands, comprising all possible random paths through the graph, after which a sequence representing the designed result was biochemically extracted from the stew.

Initially it was thought that DNA computing would be limited to the solution of combinatorial problems, but subsequent research has shown that the approach can be applied to a much wider class of digital computations. The well-known problem of combinatorial explosion leading to huge-volume DNA libraries during a complex computation may be avoided using a recursive selection approach crudely analogous to genetic algorithms. Some work has been done using plasmids and restriction endonucleases to implement a DNA-based Turing machine, using only commercially available restriction enzymes and ligases for every operation with states represented as sequences of bases around a plasmid, and Warren Smith has another Turing machine design using the chemistry of guide RNAs in *Trypanosome* kinetoplasts. Practical challenges include finding fast, efficient, and low-noise input and output techniques.

DNA computation is in theory energetically efficient, requiring only ~12 kT per ligation operation at room temperature, with massively parallel operations although each ligation reaction requires ~1 sec to complete. DNA data representation allows up to ~1 bit/nm^3 storage density. However, a practical mass storage device would probably require 10-100 nucleotide-long words to avoid ambiguities during the recall process, thus reducing maximum density to 10^7-10^8 bits/$micron^3$, though of course some of this reduction could be offset by reducing read rate requirements.

Biomechanical Computers

Components of mechanical computers could be constructed entirely of biologically-derived materials which in theory might be fabricated using synthesis pathways accessible to engineered microbiota. Drexler's mechanical nanocomputer assumes diamondoid logic rods with a Young's modulus of $E=5 \times 10^{11}$ N/m^2, but the Young's modulus for human bone hydroxyapatite and fluoroapatite crystals is almost as high, with $E \sim 1\text{-}2 \times 10^{11}$ N/m^2. Ignoring the significant problems inherent in noneutactic rod logic systems, apatite-based mechanical computer structures might be laid down using engineered osteocytes, perhaps assisted by chemotactic or contact guidance techniques. Failure strength of apatites is ~100 times poorer than for diamond, so apatite logic rod force and acceleration should be reduced by ~100 and rod vibrational energy by ~10^4, increasing minimum switching time by a factor of ~10 (to ~1 nanosec) and decreasing maximum rod speed to ~1 m/sec.

N. Seeman has constructed molecular building blocks from unusual DNA motifs, using stable-branched DNA molecules with the connectivity of a cube or a truncated octahedron. E. Winfree has proposed using arrays of DNA crossover molecules in DNA-based computing, requiring the ability to build periodic backbones with bases differing from one unit cell to another. In addition to branching topology, DNA also allows control of linking topology. DNA-based topological control has led to the construction of Borromean rings which could be used in DNA-based computing applications. In Borromean rings, the linkage between any pair of rings disappears in the absence of the third. Rings can be designed with an arbitrary number of circles; the integrity of a link could represent the truth of each of a group of logical statements.

DNA structural transitions could be used to drive nanomechanical devices via branch migration. Application of torque to a cruciform leads to the extrusion or intrusion of a cruciform. A synthetic branched junction with two opposite arms linked can relocate its branch point in response to positive ethidium-induced supercoiling, representing the first experimental step in developing DNA structural transitions that can achieve a nanomechanical result. The possible use of the B-Z transition in nanomechanical devices is being explored.

In 1998, Winfree and colleagues in Seeman's laboratory described a simple, predictable, highly precise technique for arranging DNA molecules into two-dimensional crystals. At the same time, a paper by Heath and others described a defect-tolerant computer architecture called the Teramac, a massively parallel experimental computer which apparently could operate normally with at least half of its most critical components failed. The authors claimed this architecture showed it would be "feasible to chemically synthesize individual [molecular] electronic components with less than a 100 Percent yield, assemble them into systems with appreciable uncertainty in their connectivity, and still create a powerful and reliable data communications network."

One can also imagine a network of interlocking enzyme-like conformational switches, possibly embedded in a semirigid two- or three-dimensional scaffolding of DNA or protein molecules. Enzymes may display multivalent feedback inhibition or may have multiple metal-ion binding sites influencing enzymatic action; an enzyme requiring two conformational changes to inhibit its action could serve as a biomechanical NAND gate if tethered to other similar enzymes in an appropriately structured protein array. Single-device error rates will be high but might be reduced to acceptable levels by employing short multiply-redundant pathways for each digital computational operation, or a sufficiently parallel architecture. Site-directed mutagenesis optimization of antibodies has shown that natural proteins can be made significantly stiffer and more stable by the mutation of just a few amino acids. Systems of reciprocating actomyosin molecular motors might serve as interdigitating components of a push-pull mechanical logic architecture resembling rod logic or a three-dimensional loom. In early 1999, Viola Vogel and colleagues discovered, via computer simulation, a tension-activated fully-reversible biomechanical switch comprised of a single strand of fibronectin protein – like untying a shoelace, a slight mechanical tug unravels a folded segment, switching off the protein's biochemical activity.

Mechanical coupling of intracellular Ca^{++} ion release channels allows coordinated voltage gating across the surface membrane of the sarcoplasmic reticulum inside muscle cells. R. Bradbury [personal communication, 1999] suggests that site-directed mutagenesis could produce a family of proteins that could be activated by the release of various mono- (e.g., K^{+}), di- (e.g., Ca^{++}), and tri-valent (e.g., Al^{+++}) ions. This would allow multivalued logic (e.g., K=1, Ca=2, Al=4), or increased parallelism or bandwidth (e.g., Ca=!(K & K), Al=!(Ca & Ca), etc.); once a protein exists that can be activated by a single ion, then varying its architecture to allow substitution of an ion of different size or charge is a relatively straightforward engineering exercise. A biomechanical Turing machine design has also been reported by Shapiro.

Organic and Bioelectronic Computers

Many organic materials may be useful in electronic computing. Electrical conduction in synthetic organic molecules is well known and already exploited in liquid-crystal displays of laptop computers. Doped polyacetylene is a conducting plastic with a carbon backbone; undoped, it is an insulator. There are many other known families of conducting polymers such as polyaniline, which includes nitrogen atoms in the backbone, and polythiophene, which includes sulfur. Three-dimensional interconnect arrays of organic conducting polymers have been fabricated; conductivity is higher along a chain than between chains, so sheets of oriented chains can produce relative conductivity anisotropies as high as 100-1000. An all-plastic transistor has been built. Organic transistors have been used as the active semiconducting element in thin-film transistors fabricated with organic-material thicknesses ranging from 5-150 nm, and organic microscale transistors and other organic devices may exploit bulk-effect electron transport just like silicon-based semiconductor devices. Conjugated polymer poly(1,6-heptadiester) was employed to make an optical correlator with a peak processing rate of 3×10^{16} operations/sec. Spin-transition polymers could serve as thermal sensors or memory devices with the theoretical ability to store one bit in a ~4 nm cube (~2×10^7 bits/micron^3 storage density) with ~GHz addressing speeds at room temperature. Other organic polymers possibly useful in logic circuits have been investigated.

Natural biological materials may also be useful in electronic computing, particularly because of the ability of biological materials to self-assemble. Biomolecular electronics is a subfield of molecular electronics that investigates the use of native as well as modified biological molecules in place of the organic molecules synthesized in the laboratory. In theory, a three-dimensional DNA or protein scaffolding could be self-assembled, with bacteriorhodopsin, ferritin and magnetite, or related bioelectromagnetic molecules inserted into precise locations within the structure to produce a crystal lattice bioelectronic or biooptoelectronic nanocomputer. DNA already has been used as an atomic scaffold upon which to build silver wires ~100 nm in diameter and has been decorated with fullerenes. Self-assembly can generate membrane-based and biotubule-based devices, and site-directed mutagenesis is a valuable tool for high-resolution protein device engineering. Nucleotide sequence-specific electron transfer between metallic electron donor and acceptor complexes covalently or noncovalently intercalated into strands of B-DNA has been demonstrated.

Perhaps the best-known bioelectronic (and biochemical) computer is the neuron cell and its congeries – the ganglia, nerve trunks, and brains. In 1997, David Stenger and James Hickman were attempting to culture and link together neuronal cells to build a bioelectronic cell-based sensor "computer" for performing complex pattern-recognition tasks. In early 1999, W.L. Ditto at Georgia Tech announced the creation of a biological computer comprised entirely of leech neurons, that was capable of perfoming simple sums. We can certainly imagine large numbers of cultured neurons arranged in artificial three-dimensional spatial patterns to produce a synthetic bioneural computer, but such a computer would operate at only ~KHz frequencies and would be very energy inefficient since each neuron consumes ~10^{10} kT per discharge – thus offering few advantages.

TEMPERATURE EFFECTS ON MEDICAL NANOROBOTS

Will medical nanorobots retain functionality at unusually high or low operating temperatures? This is an important question for nanodevices at work in human limbs that may be subjected to hot scalding, combustion, explosion, or other burn traumas, or, at the cold temperature extreme, to severe frostbite or exposure of the extremities whether in space or in arctic conditions, accidental ice burial after an avalanche, or in situations requiring the repair and resuscitation of cold-vitrified or cryogenically-preserved tissues, organs, or whole organisms.

A complete review of the many effects of environmental temperature on the operations of medical nanodevices is beyond the scope of this text. This Section can only briefly mention a few of the many design issues that may arise if nanorobot operations are contemplated significantly above or below normal human body temperatures.

Dimensional Stability and Strength

At high applied stress, covalent bonds cleave much more readily at higher temperatures. Example, at an applied stress of 8 nN/bond a C=C double bond cleaves in ~10^{27} sec at 0 K, ~10^{4} sec at 300 K, and ~10^{-3} sec at 500 K. The positional uncertainty is somewhat less problematic in design – the mean classical thermal longitudinal displacement of a diamondoid logic rod varies only as ~$T^{1/2}$. The endpoint of such a rod 1 nm wide and 100 nm long displaces ~0.05 nm at 77 K (liquid N_2), ~0.10 nm at 300 K, and ~0.14 nm at 600 K (liquid lead). Mechanical elements thus become more reliable at lower temperature – e.g., the probability of error in a force sensor scales as ~exp(1/T), so a 1 Percent error in a sensor at 310 K increases to a 10 Percent error at 600 K, but falls to just 10^{-6} Percent error at 77 K.

Most materials contract when cold and expand when hot. Thus a 1000-nm long diamondoid rod at 310 K contracts to ~999 nm at 77 K and expands to 1001 nm at 600 K. However, the coefficients of thermal expansion and various other thermophysical parameters are themselves temperature-dependent. The coefficients of volumetric thermal expansion at ~298 K (~25°C) are $3.5\times10^{-6}\ K^{-1}$ for diamond, $15.6 \times 10^{-6}\ K^{-1}$ for sapphire, $1.2\times10^{-6}\ K^{-1}$ for vitreous silica and $36 \times 10^{-6}\ K^{-1}$ for crystalline silica or quartz.

Young's modulus (modulus of elasticity) and the other moduli are also temperature-sensitive, especially and most obviously near phase changes — e.g., sapphire melts at 2310 K but "softens" at 2070 K. At ~1773 K, diamond plastically deforms at just ~1 atm pressure, but >60,000 atm are required to deform diamond at ~1300 K. At the cold extreme, anyone who has seen the hammering of a nail using a piece of banana cooled in liquid N_2 has witnessed the temperature sensitivity of materials strength.

Viscosity and Locomotion in Ice

The viscosity of liquids generally declines with rising temperature following Andrade's formula; water is ~6 times more viscous at 273 K (near freezing) than at 373 K (near boiling). Viscosity affects internal fluid transfers but also locomotion. Natation in liquid nitrogen should require relatively less power than in water, since liquid air (at 81 K) is only one-quarter as viscous as liquid water at 310 K.

An equally relevant but far more serious challenge is locomotion through solid water ice. Just below freezing, crystalline ice viscosity is ~10^{10} kg/m-sec, requiring a 1-micron nanorobot to expend on the order of ~200,000 pW to creep forward at 1 micron/sec by viscoplastic flow in which ice crystals are deformed without breaking. Just halfway from freezing to liquid nitrogen temperature, at 164 K, viscosity has already risen to ~10^{21} kg/m-sec, roughly equivalent to solid mantle rock, and the power requirement has increased 100-billionfold, clearly prohibitive.

One solution that avoids this problem at temperatures near the melting point is baronatation, which depends upon the fact that water, almost uniquely, is less dense as a solid than a liquid (i.e., ice floats), suggesting a freezing point depression effect with increasing pressure that is visible as the short downleg from 0°C to -16°C in the phase diagram for Ice. This is confirmed experimentally by suspending two heavy weights from a wire stretched across a block of ice. The wire passes slowly through the block, the wire exerting a pressure that melts a thin layer of water ahead of it, allowing the wire to progress; as the water passes behind the wire to the lower pressure region, it refreezes, a process known as regelation. The melting ice ahead of the wire absorbs the heat of fusion while the refreezing water gives up the heat of fusion, with heat steadily transferred by the wire, hence a good conductor cuts better than a poor one. The barostatic freezing point depression constant for ice is 134 atm/°C up to ~2100 atm. By exerting a higher pressure ahead of it than behind it, a baronatating nanorobot of roughly conical geometry can progress slowly through ice that is no colder than -16°C.

Taking the heat of fusion for water ice as ΔH_{fus}=306 pJ/micron^3 (334 J/gm at 0°C) and assuming at least ~3 micron^3 of ice must be melted to allow 1 micron of forward progress, then baronatation power requirement is very conservatively estimated as $P_{baro} \sim 3\ \Delta H_{fus}\ v_{nano} \sim 900$ pW for v_{nano}=1 micron/sec. (This is energy flow, which is not necessarily energy dissipation, since most of the heat loss will come from water refreezing at the rear of the nanorobot and only the losses from finite thermal conductivity must be made up.) Below -16°C the ice would have to be heated to that temperature before melting can occur, requiring a probably prohibitive additional power dissipation $P_{heat} \sim L_{nano}\ K_t\ \Delta T$ ~22,000,000 pW to produce a ΔT=10°C warming of size L_{nano}=1 micron in ice assuming thermal conductivity of $K_t \sim$ ~2,200,000 pW/micron-°C at -24°C. Bejan and Tyvand have analysed gravity-induced pressure-melting of ice due to the passage of solid bodies having square, disklike, or cylindrical contact surfaces. The details of compression-driven phase transitions in ice are also being studied computationally using the tools of molecular dynamics.

The freezing point depression effect in which solute molecules are released ahead and recovered behind might serve as the basis for a similar drive system concept at temperatures just below the melting point.

Burrowing by progressive voids is yet another alternative that will work over a wider range of cold temperatures. The binding energy per hydrogen bond in the ice ahead is E_{HBond}=33 zJ/bond, there are two hydrogen bonds per water molecule, and $n_{water}=3\times10^{10}$ water molecules/micron^3 in ice at 273 K, so H-bond-breaking power is at most $P_{HBond} \sim 2\ E_{HBond}\ n_{water}\ L_{nano}^2\ v_{nano}=2000$ pW for a nanorobot of dimension L=1 micron and velocity $v_{nano}=1$ micron/sec. Additionally, a molecule handling device of an efficiency comparable to the telescoping manipulator arm that moves ice molecules a total distance ~10 nm to and from a conveyor device of efficiency ~10^{-6} zJ/nm per molecule running a ~1 micron course consumes $E_{transport}=100.001$ zJ/molecule, so molecule-moving power is at most $P_{move} \sim E_{transport}\ n_{water}\ L_{nano}^2\ v_{nano}=3000$ pW, hence total motive power for a 1-micron ice-burrowing nanorobot moving at 1 micron/sec is at most ~5000 pW. Moving ice in small chunks may be another energy-saving alternative.

Solubility and Solvents

Gas solubility always declines, while the solubility of most solids rises, with higher temperature. Liquid solutes behave as solids in this regard, but note that unit volumes of miscible fluids are not precisely additive. If 1 liter of ethanol is added to 1 liter of water, the result is only 1.93 liters of solution.

A solution of a nonvolatile nonelectrolyte solute has a boiling point T_{boil} that is elevated above the boiling point of pure solvent by $\Delta T_{boil}=n\ k_b\ Ml$, where Ml is the molality of the solution, k_b is the boiling point constant for the solvent, and n=1. The presence of solute also slightly lowers solvent freezing point T_{freeze} by $\Delta T_{freeze}=n\ k_f\ Ml$, where k_f is the freezing point constant for the solvent. Both effects are a direct consequence of the lowering of solvent vapour pressure by a solute. The constants may be estimated as:

$$k_f = \frac{RT_{freeze}^2}{1000\,\Delta H_{fus}}\ (C/molal)$$

$$k_b = \frac{RT_{boil}^2 MW}{1000\,\Delta H_{vap}}\ (C/molal)$$

where R=8.31 J/mole-K (universal gas constant), MW is molecular weight in gm/mole or daltons, T_{freeze} and T_{boil} in

K, and ΔH_{fus} and ΔH_{vap} are the heats of fusion and vapourization, respectively.

In the case of electrolyte solutes (e.g., KCl) in dilute solutions (e.g., Ml <~ 0.01 molal), n is approximately the moles of ions per mole of solute. In more concentrated solutions, experimental values of n are slightly lower, as explained by Debye-Huckel theory due to ion-ion and ion-solvent interactions. With KCl in water, n=1.94 at Ml=0.01 molal but n=1.80 at Ml=0.50 molal. The most concentrated salt solution depresses the freezing point of water by ~20°C.

At still lower temperatures, solvents with much lower melting points may be employed to avoid solvent freezing. In general, substances dissolve in liquids that are chemically similar. Water and many organic compounds including glucose are soluble in ethanol, which remains a liquid to -117°C (156 K). Sodium chloride is slightly soluble in ethanol, and also in liquid ammonia which melts at -78°C (195 K). Some natural enzymes are known to retain their function in liquid ammonia, and others in supercritical carbon dioxide; artificial enzyme systems based on natural peptides, or other polymers with protein-like conformational properties, could in principle operate in cryogenic solvents such as tetrafluoromethane, carbon monoxide or liquid nitrogen. Lactate dehydrogenase enzyme found in cold water Antarctic fishes operates as fast as the related enzyme in animals with higher body temperatures, even though enzyme action normally halves for every 10°C temperature drop. The Antarctic fish enzyme compensates for the cold with modifications near the enzyme's active site that increase flexibility and mobility, in effect "greasing the hinges so that the enzyme can move more quickly.

Heat Conductivity and Capacity, and Refrigeration

Heat capacity (C_V) generally rises with temperature. The heat capacity of ice rises from 0.63 x 10^6 J/m^3-K at -196°C (liquid N_2) to 1.7 x 10^6 J/m^3-K at -30°C. Thermal conductivity (K_t) may have a more complicated relationship with temperature. The thermal conductivity of liquid water rises with temperature, from 0.561 W/m-K at 0°C to 0.681 W/m-K at 100°C. The thermal conductivity of sapphire normal to the optical or c-axis also rises from 2.3 W/m-K at 310 K to 6.0 W/m-K at 900 K, according to one source; however, in the direction parallel to the optical axis, the thermal conductivity of sapphire rises from 0.3 W/m-K at 3 K to a peak of ~200 W/m-K near 70 K, then falls to ~30 W/m-K at 310 K and ~6 W/m-K at 1000 K; conductivity for ruby falls from ~20 W/m-K at 310 K to ~6 W/m-K at 1000 K. In the case of diamond, thermal conductivity rises from ~10 W/m-K at 3 K to a peak of 12,500 W/m-K at 69 K, then falls to ~2000 W/m-K at 300 K.

Is it possible to build micron-scale refrigerators? Refrigerators serve to maintain a temperature differential between an enclosed volume and the external environment. Leaving aside vacuum isolation levitation techniques, thermal equilibration of a volume of size L by conduction requires an equilibration time of approximately:

$$tEQ \sim \frac{L^2 C_v}{K_t} \ (sec)$$

For water at 310 K, C_V=4.19 x 10^6 J/m^3-K and K_t=0.623 W/m-K, hence t_{EQ} ~ (6.7 x 10^6) L^2 (sec) for in vivo medical nanorobot refrigerators. Thus a cold box 1 mm wide requires t_{EQ} ~ 7 sec, but a 1 micron^3 box equilibrates in only ~7 microsec. Diamond is far more conductive; a 1 micron^3 cold box embedded in a surplus of diamondoid structure (C_V=1.8 x 10^6 J/m^3-K) inside a nanorobot has a t_{EQ} ~ (900) L^2=0.9 nanosec equilibration time (giving a ~10^{11} kelvin/sec cool rate, assuming DT ~ 100 K). Thus in order to avoid thermal re-equilibration with the surroundings, a ~1 micron^3 refrigeration mechanism must circulate working coolant fluid at a velocity v_{fluid} ~ L / t_{EQ} ~ 1000 m/sec, very near the speed of sound in most fluids, hence sub-ambient refrigerators smaller than ~1 micron^3 are not feasible using this method.

Drexler proposes a working fluid consisting of encapsulated submicron water ice particles with surface structures preventing aggregation and flexible enough to allow repeated expand/contract cycles as the contained ice alternately freezes and thaws, combined with a low-viscosity, low-melting point carrier such as a light hydrocarbon. The heat absorbed by a substance that melts at constant temperature is the heat of fusion; the much larger heat required to boil a liquid is the heat of vapourization; the heat of sublimation is simply the sum of the two. Phase changes provide the most efficient cooling – ice absorbs 306 pJ/micron^3 of heat when it melts at 0°C, and water absorbs 2170 pJ/micron^3 when it boils at 100°C, but water at 37°C absorbs only 42 pJ/micron^3 when it warms by 10°C. The exact temperature of a reversible phase transition in a refrigerant working fluid can be precisely controlled by judicious selection of operating pressures and fluid materials. Thermal-driven phase-change microactuators have been tested.

Many other possible refrigeration technologies are known but have yet to be investigated in the context of nanorobot refrigeration, including "magnetic cooling" by adiabatic demagnetization or magnetocaloric refrigerators, Seebeck effect or Peltier effect electronic cooling, thermoacoustic refrigeration, optical refrigerators, chemomechanical turbines operated in reverse, heat of solvation cooling mediated by molecular sorting rotors heat of allotropic-transition cooling and acoustic, polymeric, or mechanical prevention of ice crystallization during supercooling of working fluid.

TEMPERATURE-DEPENDENT PROPERTIES

There are a great many temperature-dependent materials properties of nanomedical relevance, but there is only space here to mention just a few:

1. *Denaturation and Combustion*: Protein denaturation and reduced receptor-ligand fidelity may occur at temperatures as low as 50-100°C. It is true that not much happens to a block of diamond dropped into boiling water, but the maximum combustion point for diamond in air is ~800°C, and carbon nanotubes start to burn in air at ~700°C. At low temperatures, receptor-ligand binding may occur with greater

fidelity but more slowly, and there are various unusual biological effects that occur at low temperature – the lens of the human eye becomes opaque when cooled to below freezing.

2. *Speed of Sound*: Acoustic waves travel at different speeds in cold and hot media, potentially affecting medical nanorobot sensing, energy transmission, communication and navigation. In general, the speed of sound (v_{sound}) in liquids depends upon the adiabatic bulk modulus B and density r, both of which are temperature-dependent, as $v_{sound}=(B/r)^{1/2}$. The temperature dependence of the speed of sound in pure water at 1 atm has been carefully studied and is approximated fairly well by:

$$v_{sound} \sim 1557-(0.0245)(347-T)^2$$

where T is temperature in K. Thus the speed of sound increases with rising temperature up to a peak at 347 K (74°C), then decreases thereafter. For practically all other liquids, v_{sound} decreases with rising temperature over the entire range in which the material remains a liquid. The speed of sound in water-ice just below the freezing point, is ~1000 m/sec. The speed of sound in dry air at 1 atm increases with rising temperature and is approximated by:

$$v_{sound} \sim 332-\left[(1+0.003366)(T-273)\right]^{1/2}$$

3. *Energy Absorption*: Acoustic absorption and attenuation coefficients change with temperature, affecting the efficiency of acoustic power transfer. Absorption per unit of radio frequency (rf) energy in tissue during diathermic heating also varies with temperature.

4. *Surface Tension*: The surface tension of liquids at the airliquid interface tends to decline as temperature rises, falling to zero at the boiling point. For instance, the air-liquid surface tension for a 48 Percent

volumetric ethanol-water mixture is 30.10 $\times 10^{-3}$ N/m at 20°C but 28.93 x 10^{-3} N/m at 40°C.

5. *Dielectric Constant*: The electrical properties of materials may be temperature-dependent. The dielectric constant of water declines with rising temperature and is crudely approximated by:

   $$k_{water} \sim 80 - 0.4(T - 293)$$

   for temperatures in kelvins from T=273-373 K (0-100°C), for rf frequencies up to 100 MHz, and at 1 atm pressure. The dielectric constant of ice at 0°C is virtually the same as that of water, but decreases rapidly with decreasing temperatures below 0°C, and with increasing frequency; by 0.1 MHz, $k_{ice} \sim$ 2-4 with little influence of temperature. Relative permittivity decreases by a large factor for many other substances as they change state at the freezing point; K falls from 35 for liquid nitrobenzene at 279 K to 3 for solid nitrobenzene at 279 K. In general, nonpolar liquids have a small dielectric constant that is nearly independent of the temperature, whereas polar liquids have a larger value that declines rapidly as temperature rises.